Praise for Climate Connection: American Student Voices

"What a refreshing idea it is to have the next generation of leaders engage in researching climate change and the consequences to local, national and global ecosystems. It is wonderful to see these young people taking on these intense issues and reflecting on the inactions of our current leaders. I certainly hope that these leaders-in-the-making will continue to take the climate crisis seriously and endeavor to make the hard decisions that should have been addressed decades ago."

Wil Hershberger
The Songs of Insects
Hedgesville, WV

"This work is a timely, compelling, and vitally important read."

Lee Temple
Founder
Primamundi.com

"Any time you can get students from diverse backgrounds to work together toward a common good, you have a winner. If that good is wise stewardship of our shared planet, better yet. To aspire to unravel the issue of climate change and its social consequences is a tall order. One should beware of simplistic solutions to complex problems. A discerning, compassionate mind will be required to find interdisciplinary, real-world solutions to the problems that climate change present."

John Cancalosi
nature-photography.us
Germantown, MD

Climate Connection: American Student Voices

Pamela S. Ellis, Ph.D.

Editor

Books Motivate Press

The Climate Connection: American Student Voices book may be purchased for educational or sales promotional use. For information, please contact Books Motivate Foundation, Inc. at (304) 404-4150 or business@booksmotivate.org.

Citation:

Books Motivate Foundation. (2022). Climate Connection: American Student Voices (P. Ellis, Ed.). Books Motivate Press.

Published by Books Motivate Press 08/31/2022

Books Motivate Foundation, Inc.

101 W. Washington Street, #454

Charles Town, WV 25414

www.booksmotivate.org

Phone: 1 (304) 404-4150

ISBN: 9798218030261 (Paperback)

ISBN: 9798218030278 (Hardback)

Library of Congress Control Number: 2022943869

Books Motivate Foundation, Inc.

Climate Connection: American Student Voices/ Books Motivate Foundation, Inc.

1. Climate Science 2. Education 3. Public Policy

Preface

As this book prepared for publication in July 2022, the United States news media announced that Senator Joe Manchin, West Virginia decided to support climate legislation and all U.S. Democratic Senators except for possibly Senator Kyrsten Sinema, Arizona. All U.S. Republican Senators opposed climate legislation. Our future depends on our ability to act individually and collectively for the common climate good. The necessity to find agreement and strategy to achieve climate objectives quickly is deftly articulated by Justin Guay, Director Global Climate Strategy Sunrise Project:

> The largest systemic risk of them all, climate change, is driven by reckless investments in fossil fuels, exactly the kind of speculative activities Dodd-Frank was designed to bring to a halt in order to prevent a repeat of the 2008 financial crisis. That means Dodd-Frank gives the Biden administration the power to inhibit or prohibit investments in fossil fuels — a power that could be critical for achieving his pledge of delivering a carbon-free power sector by 2035.[1]

---

[1] Guay, J. (2020, December 30). The most important climate legislation has already passed. Greentech Media. https://www.greentechmedia.com/articles/read/the-most-important-climate-legislation-has-already-passed

Introduction

The National Climate Essay Student Competition sponsored by Books Motivate Foundation was the impetus that resulted in student-led climate inquiry for recently concerned readers, and often-jaded generations of weary leaders and followers, with the inspiration and hope necessary to access a sustainable climate future. American high school students were asked to provide written responses:

1. *What is an example of an important climate science concern in your home state?*

2. *Why is this example important?*

3. *How effective have scientists, policymakers, and community members been in addressing this concern?*

4. *What more needs to be done, and by whom?*

5. *Working with those concerned, how can you make a difference now, and possibly in the future?*

Twenty-three selected climate essays from students are featured as they grappled with sense making to use their power to exercise their research skills and personal agency to make a difference within the climate and education policy world landscape.

The first 2019 National Climate Student Essayist, Jasmine DeMaria, a West Virginia resident came from a state that currently has only one in total International Baccalaureate (IB) School in comparison to its neighboring states of Virginia (73), Maryland (73), Ohio (46), Pennsylvania (32), and Kentucky (8). Avid proponents of IB especially appreciate the assessment of objectives that:

- encourage students of all ages to think critically and challenge assumptions

- develop independently of government and national systems, incorporating quality practice from research and our global community of schools

- encourage students of all ages to consider both local and global contexts

- develop multilingual students.[2]

---

[2] International Baccalaureate. (2022, June 20) Why the IB is different.
https://www.ibo.org/benefits/why-the-ib-is-different/

The mission of education to broaden the horizon of every student and their generation to think and act locally, nationally, and internationally is paramount for human and ecological survival. Active rather than passive learning creates a pathway for attainment of crucial personal and civic goals. Each one of us must decide if the artificial limitations that we are offered are sufficient to contain our audacity to plan for a different life expectation for ourselves and for planet Earth. This book showcases the wisdom of the next generation of world leaders, high school students who care deeply about their present-day and future climate quality of life prospects.

Pamela S. Ellis, Ph.D.
Editor, Executive Director
Books Motivate Foundation
Charles Town, WV

Contents

Photo Credit: Grist / Smith Collection / Gado / Getty Images.

I.

Shoreline Beaches * Habitable Land * Wildfires

We broke down what climate change will do, region by region. Grist. (2021, July 13).
    https://grist.org/cities/we-broke-down-what-climate-change-will-do-region-by-region/

Wave Goodbye to California's Beaches

Arun Brahma

Piedmont High School, Piedmont, CA

In California, when people think of climate change, they often think of the horrible forest fires that have plagued the entire state each fall for the last decade, causing deaths, billions of dollars of damage to homes and businesses, school closures, and widespread power outages. However, with all the attention given to the fires, people often overlook another climate change problem affecting California: beach erosion. The beaches are among the most important lifestyle, and tourism draws for coastal California. I fondly remember learning how to surf off of the beaches in San Diego as a child but being scared of going to surf at the Opal Cliffs in Santa Cruz.

Dr. Stephen Leatherman, a Professor and Director of the Laboratory for Coastal Research at Florida International University, said that "between 80 and 90 percent of the sandy beaches along America's coastlines have been eroding for decades" (2008). Erosion is a particular problem for California, which could lose two-thirds of its beaches as a result by the year 2100, according to the United States Geological Survey. As sea levels rise and coastal storms become more frequent and more severe, beaches in California are beginning to be overtaken by seawater, and sand is being shifted to deeper parts of the ocean. At the current pace of erosion, famous beaches, like Pacifica in Northern California and Malibu beach in Southern California, will not exist within my lifetime. Erosion would leave the state with cliffs looking down into slowly rising water.

This erosion also threatens infrastructure on the coast. Houses, roads, and other projects are threatened by flooding in coastal areas with sea-level rise. As the beaches begin to be covered with water, the infrastructure loses the barrier that the beaches provide to the ocean. For example, in 1930, there were twenty-one houses perched on a bluff above the ocean at Gleason Beach. Currently, only four remain as the cliff continues to erode by one foot per year. These remaining houses sit slightly closer to the ocean than a stretch of Highway 1 that the California Department of Transportation (Caltrans) is spending $26 million to move 850 feet inland. In the 21st century, it is estimated that 600,000 Californians could be forced to leave their homes due to extreme sea-level rise and beach erosion.

Preventing this erosion is harder than one might think. California has historically used the process of beach nourishment, which involves transporting sand from other places to areas in need, to keep the beaches alive. However, that process is unsustainable as sea levels begin to rise at a more drastic rate for two reasons: the sheer amount of sand California would need to replenish beaches is unattainable; and, if water rises enough, the surf will break into cliffs leaving no space to add sand. Normally, beaches are replenished naturally, but storm drains near the coast tend to take away all of the sediment that would normally go to the beaches.

California has also been using stone riprap, rocks placed on shorelines to protect against erosion. However, this process transforms our soft, pristine beaches into rocky, pebbly places that people shy away from rather than visit. Other solutions include beach armoring, which builds a structure, like a seawall, to protect everything behind it, but such structures do not protect the beach itself. Breakwaters structures off the coast act as barriers for waves to go to the part of the beach they are protecting and are effective at regrowing the beach behind them, but that beach regrowth comes at the expense of the unprotected beaches where all of the deflected waves and currents go.

California is at risk of losing one of the quintessential parts of its identity, as millions of people flock to California beaches annually. Even for Californians who do not live in coastal communities or visit the beach frequently, this will have a profound effect because economic activity in the coastal communities most at risk generates $40 billion for the California economy every year. In addition, a significant drop in tourism revenue due to rocky or vanishing beaches would also result in over twelve million people in coastal California communities losing their jobs.

As citizens of California, we can promote the nourishment of plants whose roots trap sand, such as coastal grasses, along our beaches. In addition, we need to educate the public on the importance of avoiding behaviors that slow or prevent vegetation growth, like littering and digging holes in the sand. Of course, those solutions are all temporary and can help, but at the end of the day, we need to be more conservation-minded to prevent further global warming and sea-level rise, which are the root causes of the beach erosion problem.

Works Cited

Admin. "How to Prevent Beach Erosion." *Best Beaches Near Me*, 12 Nov. 2021, bestbeachesnearme. com/how-to-prevent-beach-erosion/. Accessed 28 May 2022.

"Beach Erosion on the California Coast." *California Beaches*, www.californiabeaches.com/california-beach-erosion/. Accessed 28 May 2022.

"California." *National Oceanic and Atmospheric Administration Office for Coastal Management*, 27 May 2022, coast.noaa.gov/states/california.html. Accessed 28 May 2022.

Coastal and Marine Hazards and Resources Program. "Disappearing Beaches: Modeling Shoreline Change in Southern California." *United States Geological Survey*, 31 May 2017, www.usgs.gov/ programs/cmhrp/news/disappearing-beaches-modeling-shoreline-change-southern-california. Accessed 28 May 2022.

"Coastal Erosion Management at Ocean Beach." *National Park Service*, 2016, www.nps.gov/goga/ learn/news/coastal-erosion-management-at-ocean-beach.htm. Accessed 28 May 2022.

Gary Griggs, Kiki Patsch "The Protection/Hardening of California's Coast: Times Are Changing," Journal of Coastal Research, 35(5), 1051-1061, (26 June 2019)

"How Do You Deal with Shoreline Erosion?" American Geosciences Institute,
www.americangeosciences.org/education/k5geosource/content/rocks/how-do-you-deal-with-shoreline-erosion. Accessed 28 May 2022.

McLaughlin, Michael. "Your Favorite California Beach May Disappear Too Soon." *HuffPost*, 27 Mar. 2017, www.huffpost.com/entry/california-beach-erosion_n_58d97a32e4b0f805b3227bfb. Accessed 28 May 2022.

"PROTECTING AND RESTORING OUR BEACHES." *LA Department of Beaches and Harbors*, beaches.lacounty.gov/beach-erosion-restoration/. Accessed 28 May 2022.

"Shoreline Erosion Control & Public Beach Restoration." *California Government Division of Boating and Waterways*, dbw.parks.ca.gov/?page_id=28766. Accessed 28 May 2022.

Smith, Lindsey J. "California's radical plan to defend homes from sea level rise: move them." *San Francisco Chronicle*, 21 Apr. 2022, www.sfchronicle.com/travel/article/California-coast-sea-level-rise-17091737.php. Accessed 28 May 2022.

"What Causes Beach Erosion?" *Scientific American*, 17 Dec. 2008, www.scientificamerican.com/article/ what-causes-beach-erosion/. Accessed 28 May 2022.

"What Threat Does Sea-Level Rise Pose to California?" *Legislative Analyst's Office*, 10 Aug. 2020, lao.ca.gov/Publications/Report/4261#:~:text=SLR%20by%202100.-,Rising%20Seas%20Threaten%20the%20California%20Coast%20in%20Numerous%20Ways,most%20commonly%20referenced%20SLR%20risk. Accessed 28 May 2022.

Wildfires & Climate Change

Ana Sanchez

Greenfield High School, Greenfield, CA

California wildfires have caused droughts and have affected climate change. These wildfires have spread throughout all of California. On the Central Coast, they tend to become a big problem. Wildfires increase air pollution in surrounding areas and can affect regional air quality. This effect of wildfires is important because it is also impacting other regions. The smoke of wildfires can range from eye and respiratory tract irritation to more serious disorders, including reduced lung function. Scientists have shown ways to reduce fire risks, including a combination of "thinning" forests with selective logging to make it more open and intentionally setting prescribed burns that mimic gentler wildfires. The community has planned to try to prevent wildfires. Unfortunately, during the California wildfires in 2016, about 669,534 acres across the state of California burned.

Another climate change concern we have in California is droughts. With more frequent droughts, there will be less water for our agricultural industry. NASA satellites supply the U.S. drought monitor data about water availability so farmers can prepare for droughts. Scientists, policymakers, and community members have all been affected by the drought data. In addition, droughts lead to more wildfires. The Department of Forestry and Fire Protection said that they are expecting fires in July and maybe in September all the way to October.

Reducing the risk of fire often involves removing vegetation that can fuel fires. Doing this would help because dry vegetation can get heated quickly and set a fire. This technique would help because it would decrease fires, although not stop them completely. We should be cautious with our forests and what we do around them because if it is a problem now, it will be an even bigger problem for our future generations. We can apply this to our state by getting to work to get rid of the dry vegetation; there should also be a time when vegetation should get covered so that it does not become dry. By covering the dry vegetation, the fires would become less common, signifying that we can focus more on the forests and how to keep them from burning. For example, you can put mulch over the vegetation. Mulching includes putting shredded leaves, brown cardboard, straw, or wood chips on them. If not, you can pull the vegetables or bury them; that way, the sun would not be able to get near them, and you do not have to worry about fires. Wildfires spread quickly. If they continue to spread, it will be hard for wild animals to survive and for our future generations, too. Removing dry vegetation and mulching can be applied to our situation because we want to reduce the frequency of wildfires, which are some of the most dangerous natural disasters.

Wildfires cause and release toxic smoke with chemicals that can cause cancer and other trauma. For example, little kids are still growing; it can cause them a lot of PTSDS. As for people with asthma, their lungs are not strong enough to handle the damage the smoke does to the lungs. Smoke also irritates the airway. Wildfires increase air pollution in surrounding areas and can affect regional air quality. The effects of smoke from wildfires can range from eye and respiratory tract irritation to more serious disorders. These health examples explain how people may be harmed when or if a wildfire happens near them. It now seems more important than ever to understand how wildfires work and their lasting implications on our health and the environment. With that information, you can now see why wildfires are affecting our environment and climate change.

Something else that may need to be done is reporting unattended fires. Everyone should correctly extinguish fires and not throw cigarettes out of a moving vehicle. These purposeful actions could help because they would prevent a fire from happening in a forest. I recommend this because although it might not seem like it will do much, it is doing a big favor for our environment. Cigarettes that have not been correctly put out can still set things on fire, so it is important to check twice before properly disposing of them.

Fireworks are one of the reasons that we have a lot of wildfires. They are commonly used on special occasions, although many individuals use them irresponsibly or illegally. We should check our surroundings to see if there is anything that can spark. Fireworks let out sparks into the air, and they could end up landing on dried-up bushes and dry leaves, causing forests to overheat and start a fire, which then affects the climate's temperature, causing climate change. By checking our surroundings, we can reduce the chance of starting a wildfire.

To sum everything up, wildfires have been causing serious effects on climate change. Many individuals lose their homes due to the fact that it costs thousands of dollars an hour to keep a helicopter flying to help put out a wildfire. L.A. county department is investing in added resources and tools to fight back the flames. If we leave campfires, there is a chance a spark of flame can go onto dry leaves of plants and cause a fire. When people go to the forest and light fireworks, it can cause the forest to heat up and light up in flames. As a high school student, I would inform people about the dangers in their daily life. Losing forests can cause us to have less oxygen and many animals to lose their habitats.

Works Cited

Ahrens, Marty. "Fireworks Fires and Injuries." *NFPA*, June 2022, https://www.nfpa.org/News-and-Research/Data-research-and-tools/US-Fire-Problem/Fireworks-fires-and-injures.

"Pacific Northwest Research Station." *Fire Effects on the Environment | Pacific Northwest Research Station | PNW - US Forest Service*, https://www.fs.usda.gov/pnw/page/fire-effects-environment.

Phillips, Carly. "How Wildfires Affect Climate Change - and Vice Versa." *The Conversation*, 3 Aug. 2022, https://theconversation.com/how-wildfires-affect-climate-change-and-vice-versa-158688.

"Wildfires." *Centers for Disease Control and Prevention*, Centers for Disease Control and Prevention, 18 June 2020, https://www.cdc.gov/climateandhealth/effects/wildfires.htm.

"Wildfires." *World Health Organization*, World Health Organization, https://www.who.int/health-topics/wildfires#tab=tab_1.

Wildfires

Aundrea Sanchez

Greenfield High School, Greenfield, CA

Shhh! BOOM! Crack! That is the sound heard as trees fall during catastrophic wildfires that continue to affect California and the rest of the world. Wildfires are uncontrollable fires that burn wildland vegetation and are a growing climate concern for many countries and one for the Central Coast here in California as well. They continue to be an urgent concern because of how destructive they have become throughout the years and how they damage wildlife habitats and people's lives by burning their homes and even causing deaths. Scientists attribute increased fire activity to at least four factors, including hot and dry summers, stronger winds, insect infestations, and human population growth near wilderness areas. The following are needed to address the ongoing problem of wildfires; a step in the right direction to reduce flammable vegetation, strengthen communities with upgrades, and encourage people to think more consciously about their environmental impact regarding how wildfires are started.

One of the main ways to deplete the creation of wildfires is to reduce flammable vegetation, which allows fires to go through forest canopies. I recommend the reduction of vegetation because it can prevent heat which is one of the main causes why wildfires happen in the first place. According to Elliott Menashe's article, "Value, Benefits and Limitations of Vegetation in Reducing Erosion," he states that vegetation helps by protecting the soil and will cause less of a chance for a fire to occur. Menashe emphasizes that "the use of vegetation in particular requires foresight and several years of monitoring and maintenance until plants are established and effective" (Menashe 1). He recommends that certain plants should be planted in areas that do not ultimately cause more fuel to a potential wildfire. Without reducing vegetation, there would be double the number of wildfires. This strategy can be done by continuing to improve forest management projects. For example, California has an effective forest management project that essentially thins or clears trees and keeps road paths clear among at-risk communities, all of which can greatly slow down a wildfire.

Humans cause more than 85% of wildfires in the U.S., and the number continues to rise yearly. However, there are many things that smaller towns and cities can do to prevent wildfires from continuing to happen. Therefore, another way to address this problem is to strengthen communities with beneficial upgrades. Designing communities better and hardening infrastructures to better prepare for a wildfire would benefit and protect communities. In his article, "How to Fight Wildfires With Science," Albert Simeoni explains that communities could reduce the risk of becoming affected by wildfires if they had better designs. Simeoni states, "Communities should contain patchworks of flammable fuels such as vegetation, houses and cars, interspersed with less flammable and nonflammable areas such as parking lots and areas cleaned of vegetation" (Simeoni 3). The reasoning behind this is that it will "decrease fire intensity" and ultimately reduce potential wildfires. Another factor in this strategy is prioritizing building materials that are not flammable and ultimately sustain different types of exposures. I would recommend this suggestion because it ultimately would benefit future generations and keep communities from the risk of fires. I believe that strengthening communities would be an effective technique used by city officials to strengthen not only their current communities but also future ones.

The final suggestion in addressing this global problem is to encourage people to start thinking consciously about their environmental impact regarding how wildfires are started. Since humans start most wildfires, it would help if more of us could become aware of how we can take small steps to prevent wildfires. In accordance with the U.S. Department of the Interior, a few major ways of preventing a wildfire includes building a campfire in open locations, dousing your campfire until it is cold, and keeping your vehicle off dry grass. I recommend that more people follow the list above whenever they are camping and just in general. This three-step approach would be effective in the long run; it would help prevent wildfires if everyone were more conscious of their actions.

In conclusion, wildfires are a growing climate concern for many countries and one for the Central Coast in California. They continue to be an urgent concern because of how destructive they have become throughout the years. Not only do wildfires cause many harmful health effects, but they destroy communities and wildlife habitats. To address the ongoing problem of wildfires, we would need to reduce flammable vegetation, strengthen communities with upgrades, and think more consciously about how our actions impact the environment regarding how wildfires are started. If the world does not address this problem, wildfires will continue to damage communities, place more people at risk of harmful health effects, and endanger more species. In my own little world, I could make a difference by following the U.S. government's suggestions to prevent wildfires whenever I am out.

Works Cited

"10 Tips to Prevent Wildfires." *U.S. Department of the Interior*, 4 May 2022, https://www.doi.gov/blog/10-tips-prevent-wildfires.

"Forest Practice." *Cal Fire Department of Forestry and Fire Protection*, California Department of Forestry and Fire Protection (CAL FIRE), https://www.fire.ca.gov/programs/resource-management/forest-practice/.

Simeoni, Albert. "How to Fight Wildfires with Science." *The Conversation*, 19 July 2022, https://theconversation.com/how-to-fight-wildfires-with-science-86103.

Climate Concerns in California

Ana Tello

California Academy of Mathematics and Science, Carson, CA

California wildfires are one of the most devastating and destructive ecological threats to the environment in the Golden State. Oftentimes, such fires occur frequently during the driest seasons of the state, summer and fall. These fires cause millions of houses to burn down to the ground, the dislocation and separation of millions of families, numerous casualties, and help cause worsened air quality. Many of these wildfires are anthropogenic and spread rapidly due to the high temperatures and low humidity. Experts say that approximately 3.2 million acres of land have been burned down in California since the start of 2020, affecting both humans and animals alike. Various individuals such as scientists, politicians, and members of the community have attempted to address the issue, but it is seemingly never enough. To help solve this pressing matter, people should work together to stop the spread of wildfires and save the land.

Scientists analyzing wildfires in California have noted that on a five-year average, 1,223,831 million acres and 1,274 structures have been destroyed. Throughout the years 2016 to 2019, the government has spent considerably less money on fire suppression methods than in previous years, indicating that the threat of wildfires has dwindled in the eyes of the local government. However, the threat persists as the number of structures affected by the fires has skyrocketed from 1,274 to 22,905. Hypothetically, imagine one day owning a business or home, and suddenly, without a moment's notice, everything you worked hard to attain and garner for yourself has been destroyed. You are forced to evacuate for your safety and leave behind the life you have worked to achieve. The number of distraught families and business owners is far too high, and the pain they feel because of such an event is unfathomable as they experience and watch their homes and communities burn down.

Scientists, policymakers, and community members alike are all trying to find multiple solutions to slow down the wildfires, prevent the destruction of land and houses, and reduce harm to wildlife. For instance, Governor Newsom committed $5.2 billion to combat this particular issue; this money will cover removing brush and dead trees and hiring inspectors for new homes. However, some criticize this method as the money does not include water storage projects. Several Californian bills could help rectify this issue, particularly SB 45 and AB 52. The former, SB 45, creates a bond with $2.2 billion to help reduce the threat of wildfires to lives, property, and natural habitats. At the same time, AB 52 updates the California Global Warming Solutions Act of 2006, which permitted the provision of more money from the Greenhouse Gas Reduction Fund for wildfire prevention.

Meanwhile, scientists have explored new methods to help prevent and combat the dangers of wildfires. Developed by Stanford researchers, an environmentally benign gel-like fluid could help common wildland fire retardants last longer in the vegetation; this material could be much more effective and inexpensive than current methods. Trials of this fluid have shown that it provides fire protection even after half an inch of rainfall. In comparison to other methods, which provide almost no fire protection, this method is significantly better.

In the same way that scientists and policymakers are trying to come up with methods to reduce the impact of wildfires or help people affected by them, communities are also forming different groups or simply informing residents on ways to help each other during those misfortunate times. An example of this is the California Fire Prevention Organization. This organization is a nonprofit, public benefit organization that is still in the works. Their main goal is to deliver life and fire safety to communities throughout California. They partner with fire departments in many communities that are at a higher risk for wildfires. An organization that exists within communities to help before fires are the Fire Safe Council. During fires, communities work together and communicate with Community Emergency Response Teams and American Red Cross local chapters.

Even though some progress has been made to lower wildfires and the risks around them, much work still has to be done to accomplish these two primary goals. First, the government is starting to require cities and counties to create fire safety standards in high-risk areas. According to an article written by MIT Technology Review, "But one of the clearest conclusions, as experts have been saying for years, is that California must begin to work with fires, not just fight them" (Temple, 2020). This shift means that the government would have to reverse fire suppression policies and rely more on prescribed burns to help clear out vegetation that builds up, which helps feed fires. Although this method will not stop or prevent the fires from occurring, it will make them less severe, easier to control, and reduce the destruction. In addition, this method of reducing vegetation would be under controlled conditions and spaced out both geographically and time-wise.

Undoubtedly, residents and victims of these wildfires would want to know how to help or how to make a change. There are several small ways we can all help to prevent fires. For example, residents can carefully dispose of smoking materials, camp responsibly, not set off fireworks or smoke bombs, clear away vegetation, and incorporate fire-resistant plants like french lavender and sage in their gardens. Citizens can also make sure to tell those close to them about these methods and make sure that the people around them are safe.

Fires will, without a doubt, happen, especially during the drier seasons of the year. However, if the government, communities, and scientists work together, we can all help make these wildfires less severe and lower the destruction that comes our way. As a state, we can all agree to do our part in lowering the risks of this issue.

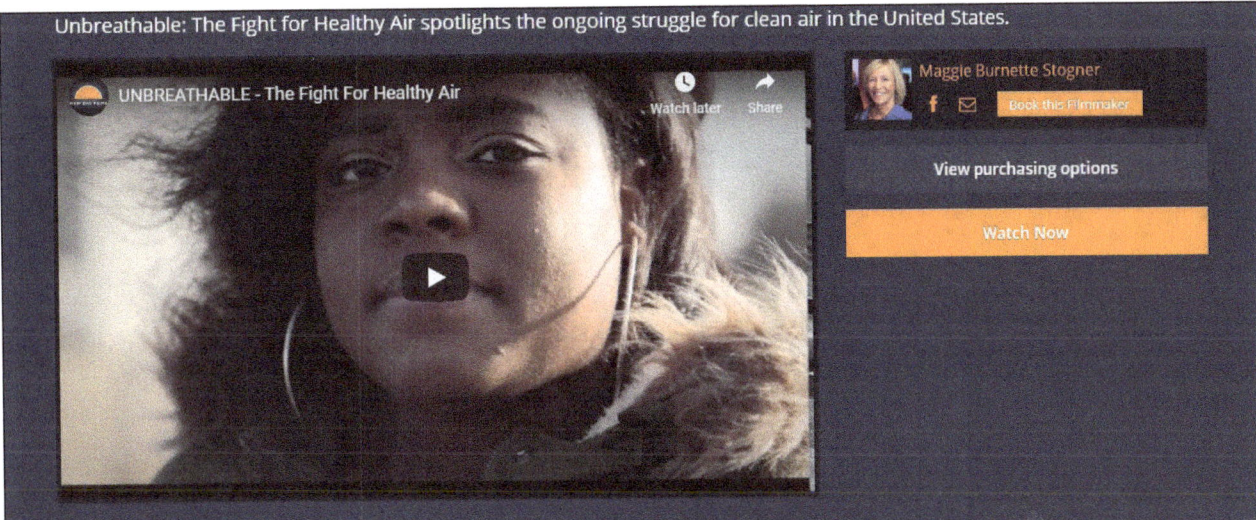

Photo Credit: New Day Films.

II.

Breathable Air * Toxic Smoke

Unbreathable: The Fight for Healthy Air | New Day Films. (2021, February 26). https://www.newday.com/films/unbreathable-the-fight-for-healthy-air

Climate

Leidy Enriquez

Greenfield High School, Greenfield, CA

Could you be the solution or the problem? One example of an important climate science concern in California is exposure to air pollution, which can lead to destroying our health and the environment. This air pollution problem is important because it increases the risk of respiratory infections and heart disease and greatly affects people who are already unwell. Also, it affects animals through stress and disease. Toxic gases and chemicals cause pollution in the air as a result of human actions. When fossil fuels are burned, nitrogen is released, which contributes to acid rain, and can cause trees to die. Scientists have addressed this problem by reducing the toxic emissions from industrial sources, and community members have been trying to lower the number of gas-powered objects.

Without continuing efforts, the air quality we breathe every morning can become more and more toxic. Air quality is important because it concerns our health. Toxic gasses also affect us because industrial emissions are another source that releases carbon monoxide and chemicals into the air. Scientists have been trying to reduce toxic emissions from industrial sources. Community members have been reducing the number of trips taken in cars and also reducing the use of fire pits to help with this problem. In addition, the government has been reducing greenhouse gas emissions through regulation initiatives and partnership programs.

Carbon use has to be reduced to help diminish air pollution, and using electric cars can help. "Research and development brought substantial improvement in battery life and lowered overall manufacturing and purchase costs" (Igini, 2022). This example explains that efforts are being made to lower the cost of electric vehicles so they can be affordable. In California, prices are always higher than in other states; if electric cars were more affordable, then they would be more common. Electric vehicles are not used much here because of the current price tag, which explains "the dearth of available electric trucks and SUVs, higher upfront costs compared to conventional vehicles." These factors need to change if we want to improve the air pollution caused by fossil fuels. The price of these vehicles can be changed based on the average income. They can increase the tax based on how much a person makes. These efforts can make Teslas more affordable. "The government is doing its part by boosting consumer tax credit." Making electric vehicles more affordable can help with toxic air pollution.

We can take public transportation to reduce fossil fuels in the air. Taking public transportation saves 4.2 billion gallons of gas which reduces the amount of fossil fuel going into the air, thus reducing the air pollution that can cause health problems. "Traveling by public transportation uses less energy and produces less pollution than comparable travel in private vehicles" (American Public Transportation Association, 2008). This evidence proves that taking public transportation reduces air pollution, which can cause health concerns like heart disease and lung cancer. "Public transportation is reducing energy consumption and harmful carbon dioxide (CO2) greenhouse gas emissions that damage the environment" (APTA, 2008). We have the option to increase or decrease the amount of fossil fuel going into the air. "Public transportation produces far fewer quantities of air pollutants." Public transportation is a solution to help air quality improve.

We can reduce indoor pollution by removing gas stoves and installing electric stoves. Gas stoves release carbon monoxide, which can be harmful to us. This indoor pollution can lead to serious health problems if we do not make a change. "But evidence has been mounting that gas stoves and ovens cause enough indoor air pollution that may be worsening kids' asthma" (Skwarecki, 2022). This evidence explains that kids with asthma can have more health problems with gas stoves in their homes; "there is evidence that suggests gas stoves are one of the pollutants that can contribute" to this problem. Gas stoves are one main reason for indoor air pollution. Change gas-powered appliances to electric. "Concerning levels of indoor air pollution … could play a larger role in driving climate change." Not only can gas stoves cause indoor air pollution they can also cause serious climate change, which can affect our ecosystem. Changing from gas to electric stoves is a simple way to help prevent indoor air pollution.

We have been causing air pollution by what we do daily. One of the main toxic chemicals in air pollution is the combination of fossil fuels and gases. The government is increasing tax credits so that electric cars can be more affordable, and electric cars have been reducing greenhouse gas emissions.

Some solutions to consider using to help air pollution are taking public transportation, changing house gas stoves to electric, and buying an electric car. I can change my gas stove to an electric one to help protect my health and my family's health. I can take public transportation. What could you do to make a change?

Works Cited

Dapcevich, Madison. "Why Natural Gas Stoves Are Harmful to Our Health and Climate." *Time*, Time, 13 May 2022, https://time.com/6176129/best-stove-for-health-environment-natural-gas-electric/.

"Environmental Benefits of Public Transit." *KCATA*, Kansas City Area Transportation Authority, https://www.kcata.org/about_kcata/entries/environmental_benefits_of_public_transit.

EPA, Environmental Protection Agency, 31 Jan. 2022, https://www.epa.gov/nutrientpollution/sources-and-solutions-fossil-fuels.

"Research and Technical Resources." *American Public Transportation Association*, https://www.apta.com/research-technical-resources/.

Skwarecki, Beth. "Are Gas Stoves Bad for Your Health?" *Lifehacker*, Lifehacker, 2 May 2022, https://lifehacker.com/are-gas-stoves-bad-for-your-health-1848869920.

"Why Electric Cars Are Better for the Environment." *Earth.Org*, 17 May 2022, https://earth.org/electric-cars-environment/.

Focus on Urban Air Pollution in California

Fiona DeFrance

Mira Costa High School, Manhattan Beach, CA

In 2020, during the Labor Day Weekend, the city of Los Angeles recorded a record spike of 185 ppb[1] of ozone pollution (*Los Angeles Times*). This event marked the most unhealthy level of ozone pollution the city had witnessed in nearly 30 years, yet most Californians received no more than an advisory to stay home for the remainder of the holiday. Each year in the United States, there are over 100,000 deaths resulting from exposure to ambient air pollution, surpassing the number of Americans killed by vehicles, Alzheimer's disease, and even gun violence (*UNC Gillings School of Global Public Health*). Many of these premature deaths could have been prevented if we were to have reduced harmful quantities of pollution to background levels, so why have we not done more to stop this death toll from climbing?

Gruesome levels of air pollution are not abnormal to natives of California, as air monitors have shown that over 90% of residents breathe in unhealthy levels of air pollutants sometime within the year (*California Air Resources Board*). Most of this air pollution derives from wildfires and photochemical smog. During an event of photochemical smog formation, nitrous oxide enters the atmosphere from sources such as vehicles and power plants and reacts with at least one VOC[2] (*National Geographic Society*). This solution of NOx[3] and VOCs are then activated by sunlight, forming tropospheric ozone and the brownish haze seen above cities (*Carleton College*).

---

[1] PPB: Refers to air pollution; parts per billion
[2] VOC: Volatile Organic Compound
[3] NOx: Nitrogen Oxides

Due to climate change, the effects of air pollution found within cities have been further exacerbated. Since photochemical smog is formed in the presence of light, this reaction process is sped up when temperatures are higher, contributing to more abundance of ground-level ozone. In addition, these higher heat temperatures transmit pollutants and diseases faster. Perpetuated by wildfires and by combustion sources within vehicles/industrial facilities, the fine particulate matter (P.M. 2.5[4]) that is released from these sources goes into our bloodstreams, which can lead to asthma, heart attacks, kidney problems, and other respiratory-related cancers (*The Guardian*). Overexposure to fine particulate matter has also been correlated to adverse birth outcomes such as low birth weight, preterm birth, neural tube defects, and cardiovascular defects (Ritz, B.). Thus, air pollution places a tremendous burden on our healthcare system and medical providers to scramble quickly to provide the necessary aid.

In efforts to combat urban pollution, policymakers have introduced the California Marketable Permits Program. This program implements economic incentives such as marketable permits, fees, and auctions of emission rights[5] to dissuade polluters from overexercising their "right" to pollute. Much like the Clean Air Act 1970[6], the California Marketable Permits Program extends the overall mission of creating safer air quality; however, it attempts to reform in a more incremental and realistic manner. Under a proposed amendment to the California Marketable Permits Program, sulfurous polluters would be required to reduce their emissions by 5% per year from their initial amount and nitrous oxide polluters by 8% per year, eventually phasing out hydrocarbon emissions entirely (Dwyer, John P.). This plan is on par with Gavin Newsom's recently passed legislation, which will require the state of California to adopt a zero-emissions policy, phasing out the sale of gas-powered cars by 2035 (Nilsen, Ella). While this is a major stepping stone towards reducing air pollution, there are still oppositions to these measures, which stem from the influence of leaders from the previous administration. In the previous administration, Congress and state legislatures had often shied away from climate change matters if they did not coincide with their political agenda. While concern for over-regulation of businesses is understandable, actively ignoring the issue of climate change and how it pertains to air pollution concerns should not come at the expense of human health or the environment.

Using the premise of the heat island effect[7], we know that those living in the hearts of cities are more susceptible to medical conditions related to issues facilitated by climate change, which has been further confirmed by hospitalization statistics in these areas (*EPA*). Unfortunately, this has political connotations as those of low socioeconomic background have historically been placed in areas in hotter parts of towns, closer to industrial activity and highways, where there are more pollutants. This is due

---

[4] P.M. 2.5: Tiny particles in the air that are two and one-half micrometers or less in width; the smaller the particulate matter, the more harmful it is.

[5] Emission credits: The government issues a certain quota of allotted pollution; it can be sold or traded to other businesses as long as the amount of pollution from both parties combined does not exceed the quota.

[6] Clean Air Act 1970: Designed to protect the public from unhealthy levels of air pollution.

[7] Heat Island Effect: Raise temperatures significantly in the middle of cities compared to their surroundings.

to racist housing policies that have left redlined[8] minorities in these undesirable conditions, hence a reason why climate change is a political issue.

The key to reducing harmful pollutants comes from participation among large businesses and the general public to value sustainable practices. In order for the California Marketable Permits Program to achieve its desired effect, facilities that pollute on a larger scale must be socially responsible to not hoard emission credits from newer firms, which may not be fully equipped yet with the technology necessary to keep within the current timeline of gradual emission reduction (Dwyer, John P.). Trading emission credits is necessary in order for there to be a productive group effort to move towards a less polluted environment. On a more localized level, we can ease the impact of air pollution by planting more trees in areas where pollution is more prevalent. Trees alleviate the effects of the greenhouse effect, which contribute to higher temperatures and improve air quality by intercepting airborne particles (*USDA Forest Service*). While we await the progress of these solutions, it is important that we do our part in conserving healthy air by adjusting our daily routine. Even adjusting habits such as turning off the lights, using properly ventilated stove tops, and getting regular car checks can make a difference in air quality, improving our health and environment one step at a time (*Cleveland Clinic*).

---

[8] Redlining: Denying services (usually financial aid) to residents on the premise of race or ethnicity.

Works Cited

"California Air Resources Board." *Health & Air Pollution | California Air Resources Board*, ww2. arb.ca.gov/resources/health-air-pollution#:~:text=A%20number%20of%20air%20pollut-ants,some%20part%20of%20the%20year.

"Can We Reduce Deaths Associated with Air Pollution?" *UNC Gillings School of Global Public Health*, 31 Oct. 2017, sph.unc.edu/cphm/can-we-reduce-deaths-associated-with-air-pollution/.

"Case Study." *Earth Exploration Toolbook*, 7 Feb. 2019, serc.carleton.edu/eet/aura/case_study.html.

"Communities and Landscapes of the Urban Northeast." *Communities and Landscapes of the Urban Northeast - Northern Research Station, USDA Forest Service*, www.nrs.fs.fed.us/units/urbanNE/.

"Doctors Alarmed by Surge in Hospital Visits as Toxic Smoke Engulfs West Coast." *The Guardian*, Guardian News and Media, 18 Sept. 2020, www.theguardian.com/world/2020/sep/18/wild-fire-smoke-health-effects-hospitals.

Dwyer, John P. "The Use of Market Incentives in Controlling Air Pollution: California's Marketable Permits Program." *Berkeley Law*, 1 Jan. 1993, lawcat.berkeley.edu/record/1114646?ln=en.

"Heat Island Effect." *EPA*, Environmental Protection Agency, www.epa.gov/heatislands#:~:text=Heat%20islands%20are%20urbanized%20areas,as%20forests%20and%20water%20bodies.

"Los Angeles Suffers Worst Smog in Almost 30 Years." *Los Angeles Times*, Los Angeles Times, 10 Sept. 2020, www.latimes.com/california/story/2020-09-10/los-angeles-had-its-worst-smog-in-26-years-during-heat-wave.

Lungs, Breathing and Allergy Team. "17 Simple Ways to Prevent Air Pollution in Your Home." *Cleveland Clinic*, Cleveland Clinic, 3 May 2021, health.clevelandclinic.org/17-simple-ways-prevent-air-pollution-home/.

Nilsen, Ella. "EPA Will Tighten Fuel Mileage Standards for Cars and Light Trucks, Replacing Looser Trump-Era Standards." *CNN*, Cable News Network, 20 Dec. 2021, www.cnn.com/2021/12/20/politics/epa-auto-emissions-standards/index.html.

Ritz, B. "Ambient Air Pollution and Risk of Birth Defects in Southern California." *American Journal of Epidemiology*, vol. 155, no. 1, 2002, pp. 17–25., doi:10.1093/aje/155.1.17.

"Smog." *National Geographic Society*, education.nationalgeographic.org/resource/smog.

Photo Credit: borjomi88

III.

Rising Temperatures * Heat Waves * Drought

Business man boat crack soil desert symbol crisis stagnation losses. Depositphotos. (n.d.).
https://depositphotos.com/360611462/stock-photo-business-man-boat-crack-soil.html

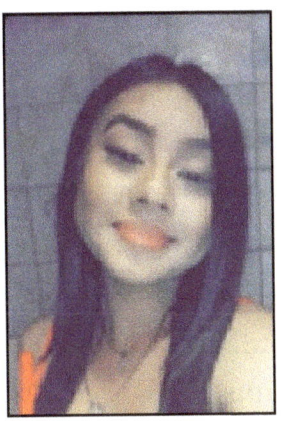

Climate Change

Mercedes De Jesus

Greenfield High School, Greenfield, CA

A climate change concern in California is heat waves. This concern is important because heat waves are becoming more common now, snow is melting, and less rain is falling in Southern California. On the news report of Denise Chow from NBC News, "Scientists published a study in the journal Nature Climate Change last month, and scientists determined that record-shattering heat-events are up to seven times more likely to occur between now and 2050, and more than 21 times more likely to occur from 2051 to 2080" (2021). Using efficient appliances in our homes can lighten the load on the electric grid during heat waves.

Trees are vital for our environment, and people need to start taking care of our trees and, if necessary, plant more. Trees help filter out air pollution, store carbon, nurture wildlife and even improve people's mental and physical health. According to the United States Environmental Protection Agency, "Trees and vegetation lower surface and air temperature by providing shade" (2022). Across the globe, hot days are getting hotter and more frequent, and heat waves are getting more dangerous when combined with high humidity. Trees provide shade and cooling through evapotranspiration. Energy Saver, an Office of the U. S. Department of Energy, states in an article, "Principles of Heating and Cooling," "If our surrounding air is cooler than our skin, the air will absorb your heat and rise. As the warmed air rises around you, cooler air moves in to take its place and absorbs more of your warmth. The faster this air moves, the cooler you feel" (n.d.). The shade that trees provide helps protect our bodies from overheating and needing to go to the hospital.

Furthermore, heat waves can burden health and emergency services and increase strain on water, energy, and transportation resulting in power shortages or even blackouts. Exposure to heat causes severe symptoms, such as exhaustion, heat strokes which can lead to death, and a condition that causes faintness as well as dry and warm skin due to the inability of the body to control elevated temperatures. Heat waves caused by human climate change have negatively affected Earth.

We need to reduce our footprint. Carbon footprint is the number of carbon compounds such as carbon dioxide emitted due to the consumption of fossil fuels from your house and other businesses. These greenhouse gases are a major contributor to sea-level rise. According to Rob Bonta, Attorney General of the State of California Department of Justice, "Approximately 85% of California's population live and work in coastal counties. The sea level along California's coasts has risen 8 inches in the past century and is projected to rise by as much as 20 to 55 inches by the end of the century" (n.d.). To avoid this, get involved and volunteer with a conservation organization near you. Volunteering is extremely helpful in improving basic health and education, tackling environmental issues, and reducing the risk of disasters.

Reducing your carbon footprint can also be done if we all support the Smog Check Program or if we walk or ride a bike when possible. Riding a bike or walking has no harmful vehicle emissions nor releases smog into the air, which is extremely helpful for our world and helps protect biodiversity. In addition, it creates less noise, less air pollution, and fewer emissions that are warming the atmosphere. In general, it improves our health and wellbeing. Opting to use our bikes a few times a week instead of our cars is one of the simplest ways to lower our environmental footprint. In addition, biking helps ensure that gasoline and antifreeze do not make their way into local waterways or the environment.

Using energy-efficient appliances and equipment is beneficial because it increases efficiency, lowers greenhouse gas (GHG) emissions and other pollutants, and decreases water use. GHGs are the leading cause of global warming and climate change. These gases are responsible for absorbing the infrared radiation that leads to holding and trapping the heat in the atmosphere. As a result, the warmth from the Earth's surface continues to increase. Everyone depends on water to live, and conserving water should be your responsibility. One of the best things to do is use home appliances that consume less water and energy. You can help reduce your carbon footprint and other GHGs by using energy-efficient home appliances since they have lower emissions of harmful gases into the environment. The "Economic Losses, Poverty, and Disasters" report states, "We live in a world where the bar for resilience is constantly being raised by human actions. The most egregious failure in this regard is the lack of political will and commitment to make serious progress on reducing greenhouse gas emissions, thus allowing climate change to play an increasingly key role in driving up disaster losses around the world for the near future" (United Nations Office for Disaster Risk Reduction, 2017).

In conclusion, heat waves are one of the biggest climate changes in California. Heat waves significantly impact society, including a rise in heat-related deaths. The California Environmental Protection Agency is working with the California Department of Insurance and the Governor's Office to plan and research to create and implement a statewide extreme heat ranking system to help people with heat waves and warn them. Growing trees and walking or bike riding can be extremely helpful for the environment. If ignored, rising sea levels could impact 1 billion people by the year 2050. Heat waves will become more frequent and severe worldwide, affecting hundreds of millions or even billions of people if we do not act. As the Earth continues to warm, crucial habitats may no longer be hospitable for certain animals or plants. An increased water temperature puts a variety of species at risk. As high school students, we could all help by consuming less meat, particularly beef. Switching over to a vegan or plant-based diet is one of the major things a student can do to help.

Works Cited

"California Releases Extreme Heat Action Plan to Protect Communities from Rising Temperatures." *Office of Governor Gavin Newsom*, Office of Governor Gavin Newsom, 28 Apr. 2022, https://www.gov.ca.gov/2022/04/28/california-releases-extreme-heat-action-plan-to-protect-communities-from-rising-temperatures/.

Chow, Denise. "Heat Wave 2021: Climate Scientists Warn about a New Normal." *NBCNews.com*, NBCUniversal News Group, 20 Aug. 2021, https://www.nbcnews.com/science/environment/heat-wave-2021-climate-scientists-warn-new-normal-rcna1664.

"Climate Change Impacts in California." *State of California Department of Justice*, Department of Justice, 20 July 2017, https://oag.ca.gov/environment/impact.

"Economic Losses, Poverty Disasters." *Prevention Web*, Centre for Research on the Epidemiology of Disasters, Emergency Events Database, & UN Office for Disaster Risk Reduction, https://www.preventionweb.net/files/61119_credeconomiclosses.pdf.

"Heatwaves." *World Health Organization*, World Health Organization, https://www.who.int/health-topics/heatwaves#tab=tab_1.

McGilvery, Kennedy. "Canada's Wildlife Is In Hot Water." *Discovery*, 28 July 2021, https://www.discovery.com/nature/canada-heat-wave--.

"Plants and Climate Change (U.S. National Park Service)." *National Parks Service*, U.S. Department of the Interior, https://www.nps.gov/articles/000/plants-climateimpact.htm.

"Principles of Heating and Cooling." *Energy.gov*, US Department of Energy, https://www.energy.gov/energysaver/principles-heating-and-cooling#:~:text=Convection%20occurs%20when%20heat%20is,moves%2C%20the%20cooler%20you%20feel.

"Using Trees and Vegetation to Reduce Heat Islands." *EPA*, Environmental Protection Agency, https://www.epa.gov/heatislands/using-trees-and-vegetation-reduce-heat-islands#:~:text=Trees%20and%20other%20plants%20help,to%20reduce%20urban%20heat%20islands.&text=Trees%20and%20vegetation%20lower%20surface,providing%20shade%20and%20through%20evapotranspiration.

Climate Change in California

Storm Medrano

Greenfield High School, Greenfield, CA

"California's drought affects everyone in the state, from farmers to fishermen, business owners to suburban residents, and everyone has a role to play in using precious water resources as wisely and efficiently as possible," a quote about California's drought by Frances Beinecke. In my home state, one of the biggest climate issues is drought. Drought is such an important topic because it has a much bigger impact on all of us than you would think; it affects our cities, California's economy, our agricultural production, our wildlife, and causes wildfires. When I was in elementary school, California was in a deep drought, and my school brought guest speakers to come in and speak to us about water conservation. Governor of California, Gavin Newsom, has asked California citizens and businesses to cut their water use by 15%. Newsom has invested in our drought situations now to build water resilience for the future and to increase water conservation outreach and education. To solve a drought that is this extreme, it will take everyone's effort to prevent as much water waste as possible.

When watering your lawns, water should be used as efficiently as possible. Efficiency means only watering your lawn as necessary and using the appropriate amount of water for your lawn. According to the University of California, Division of Agriculture and Natural Resources, lawns are estimated to use about 3.5 to 5% of the total statewide water use, so by cutting back on lawn watering habits, lots of water can be saved. These lawn restrictions have become a mandate in many areas in California due to the extreme drought. Those who fail to comply with these mandates can face fines of up to $2,000, according to Metropolitan. Some alternatives to traditional lawns include Thyme, Kurapia, and Dymandia; these lawn alternatives are drought-friendly and need less water maintenance, saving water overall.

Residents of California need to avoid the over usage of water. Water is needed for everything, so there is no way to stop water usage completely; however, there are ways to cut back on how much water is wasted. Government officials and policymakers all suggest simple alternatives when it comes to everyday activities to prevent water waste, such as taking shorter showers, fixing leaky faucets and toilets around your house, or even just turning the faucet off when it is not needed; doing these simple actions save so much water. An article, "Five Ways to Take Shorter Showers and Save Water," written by Angela Tague states, "If everyone in the United States—all 320 million (about 10 years) of us—reduced their average shower time by one minute each time, it could save a whopping 165 billion gallons (about 624592650000 L) of water" (n.d.). No one is saying not to shower to save water; however, not having a 45-minute concert in the shower is one way you can help contribute to stopping drought.

Change starts with encouragement. By spreading the word about cutting back on water usage to friends and family, you are making a difference in the community. For example, if you spend the night at your friend's house and notice that when washing their face, they do not turn off the water when it is not needed, they leave the water running while brushing their teeth, and take hour-long showers, do not be embarrassed to tell them to do better, for your future, and the future of California as a whole. The findings suggest that cutting back on unnecessary water usage adds up, and every drop is important.

Works Cited

"10 Ways Farmers Are Saving Water." *CUESA*, 15 Aug. 2014, https://cuesa.org/article/10-ways-farmers-are-saving-water.

"5 Ways to Take Shorter Showers and Save Water." *5 Ways to Take Shorter Showers and Save Water*, https://www.tomsofmaine.com/good-matters/thinking-sustainably/five-ways-to-take-shorter-showers-and-save-water.

"As Western Drought Worsens, Governor Newsom Moves to Bolster Regional Conservation Efforts." *Office of Governor Gavin Newsom*, 31 Mar. 2022, https://www.gov.ca.gov/2022/03/28/as-western-drought-worsens-governor-newsom-moves-to-bolster-regional-conservation-efforts/.

Avalos, Gina. "Central Coast Cities Seeing Impacts of California Drought." *KSBY*, KSBY, 22 Apr. 2021, https://www.ksby.com/news/local-news/central-coast-cities-seeing-impacts-of-california-drought.

California, State of. "Newsom Administration Boosts State Funding for Drought Emergency." *Office of Governor Gavin Newsom*, Office of Governor Gavin Newsom, 14 Mar. 2022, https://www.gov.ca.gov/2022/03/13/newsom-administration-boosts-state-funding-for-drought-emergency/.

"Drought Impacts on California Crops." *Caclimatehub.ucdavis.edu*, United States Department of Agriculture, https://caclimatehub.ucdavis.edu/wp-content/uploads/sites/320/2016/03/factsheet3_crops.pdf.

"Drought Response." *Association of California Water Agencies*, https://www.acwa.com/drought-response/.

Gabrielle, Vincent. "Why Planting Trees during a Drought Is a Great Idea!" *Our City Forest*, Our City Forest, 22 Oct. 2015, https://www.ourcityforest.org/blog/2015/10/21/why-planting-trees-in-drought-isnt-a-bad-idea.

Ilstedt, Ulrik et al. "Intermediate Tree Cover Can Maximize Groundwater Recharge in the Seasonally Dry Tropics." *Nature News*, Nature Publishing Group, 24 Feb. 2016, https://www.nature.com/articles/srep21930.

"Irrigation & Water Use." *USDA ERS - Irrigation & Water Use*, United States Department of Agriculture, https://www.ers.usda.gov/topics/farm-practices-management/irrigation-water-use/.

Karlamangla, Soumya. "Why Californians Have Been Saving Less Water in 2022." *The New York Times*, The New York Times, 25 Mar. 2022, https://www.nytimes.com/2022/03/25/us/water-conservation-california.html.

Larsen, Sara. "Why Should Every American Care about the California Drought?" *American Rivers*, 11 Jan. 2018, https://www.americanrivers.org/2015/05/why-should-every-american-care-about-the-california-drought/?gclid=EAIaIQobChMI7ejD9-y09wIV0j-tBh2FLgBZEAAYAyAAEgI45_D_BwE.

Matthews, Kayla. "8 Ways to Reduce Your Water Waste." *Green Living Journal*, 4 Jan. 2022, https://www.greenlivingpdx.com/8-ways-to-reduce-water-waste/.

"Solving Drought." *Environment California*, Environment California, https://environmentcalifornia.org/page/cae/solving-drought.

Stop Water Waste the Average Person Unknowingly Wastes. Washington State Department of Health, Apr. 2020, https://doh.wa.gov/sites/default/files/legacy/Documents/Pubs/331-450.pdf.

Train, Rob, and Agencias. "Drought in California: Measures, Restrictions and How Long Will They Last?" *Diario AS*, Diario AS, 28 Apr. 2022, https://en.as.com/latest_news/drought-in-california-measures-restrictions-and-how-long-will-they-last-n/.

Vidon, Tamar Shiloh. "Planting Forests Can Increase Rainfall and Help Fight Drought in Europe, Study Shows." *France 24*, France 24, 9 July 2021, https://www.france24.com/en/europe/20210709-planting-forests-can-increase-rainfall-and-help-fight-drought-in-europe-study-shows.

Vyas, Kashyap. "8 Innovative Drought Solutions That We Can Count 0N." *RIPE*, 16 June 2019, https://ripe.illinois.edu/news/8-innovative-drought-solutions-we-can-count-0n.

Weeden, Meaghan. "1.2 Million Trees Planted in California for Forest Fire Restoration in 2021." *One Tree Planted*, One Tree Planted, 5 Nov. 2021, https://onetreeplanted.org/blogs/stories/california-forest-fire-restoration-2021.

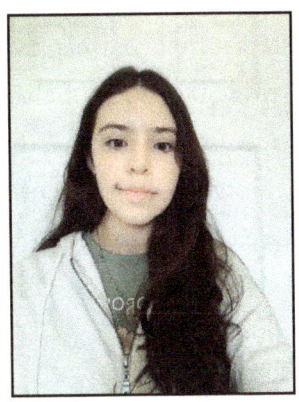

The Impact of Drought

Quetzalli Naomi Gonzalez

Greenfield High School, Greenfield, CA

Drought is "an unusual water deficit that generates severe impacts on the society that suffers from it, altering the normal development of its collective life" (United States Geological Survey, 2019). Severe drought conditions can negatively affect air quality and cause wildfires. Water is a very important part of many of our activities. Drought can affect water levels and flows, resulting in reduced river transportation access, and economic impacts are associated with this disruption. What scientists, policymakers, and the community are doing to address drought is taking significant mitigation measures, such as harvesting water, protecting water sources from contamination, and monitoring communication response plans. It is necessary to highlight that drought can cause the scarcity of drinking water. According to the United Nations Environment Program, the most effective strategies to address the problem of drought are collecting rainwater, collecting water from the air, and changing water use habits.

First, rainwater harvesting needs to be implemented more because "collecting rainwater provides urban and rural areas with an efficient option to store rainwater and then reuse it in times of drought," (Vyas, 2019). This system can be used in simple and effective methods to attend to those not requiring drinking quality, reducing more than "40% of water consumption of water in a house" (Environmental Impacts, 2015). The Center for Disease Control also mentions that "one fifth of California's electricity production is consumed by pumping and processing water" (CDC, 2021). This stat shows that rainwater harvesting should be applied in California because it can reduce our carbon footprint and conserve water resources. Rainwater harvesting means collecting water on the roofs of buildings and then storing it underground for later use. According to the United Nations study published in 2019, if just 326 people collected 1,000 gallons of rainwater in their barrels, that would leave 1 acre or 326,000 gallons of water in our reservoirs, which can provide 60 gallons of clean water for drinking and bathing, water for 5,433 people. The National Oceanic and Atmospheric Administration released a study confirming that for "most Californians, harvesting rainwater is a potential drought mitigation solution" (Olson, 2021) because homeowners are more aware of their drinking water usage, leading to better water conservation. Therefore, it is beneficial to consider collecting rainwater in homes and treating water as a precious resource that must be conserved.

Second, extracting water from the atmosphere is also a helpful solution to drought. Atmospheric water harvesting "is the capture and collection of water that is present in the air, either as vapor or small water droplets" (LaPotin, 2019). Water vapor is removed from the air in two main ways, precipitation and condensation. This technical solution is useful for drought because it can provide water in arid regions, especially in inland areas where the rural population needs water sources. Researchers at the University of California, Berkeley, have improved condensed water for arid places. In addition, atmospheric water harvesting is a "zero-energy solution for harvesting water from the atmosphere throughout the 24-hour daily cycle," (Sanders, 2018). The two main advantages of this system are that it can collect water "without the use of electricity and the ability to produce in desert climates," according to the Berkeley researchers. The University of California, Berkeley team hopes to scale up atmospheric water generators that can be applied in California so that they can produce many liters of water in a short time in a dry location. During a test conducted in Arizona by researchers at the University of California, Berkeley, they collected 0.7 liters per kilogram of absorbent per day, almost three cups of clean, pure water. This example means that the harvester can produce with extremely low relative humidity and temperatures above 80 degrees Fahrenheit. Atmospheric water harvesting helps address geographic and climatic issues and is independent of other water sources and utilities. "There is more than six times the amount of water in the form of vapor around our planet than there is in all the rivers at any given time," explains David Hertz, co-founder in California. This quote shows that atmospheric water harvesting has become a promising strategy.

Finally, changing our habits of how we use water and reducing daily usage can be a powerful way to prevent drought. The Alliance for Water Efficiency analyzed studies from various parts of the country to determine whether water restrictions were an effective tool for maintaining water supplies during shortages. They found that clear messaging must be part of the water restriction strategy. California water officials have imposed new drought rules for cities and towns across the state. These regulations, adopted by the State Water Resources Control Board, "prohibit over-watering lawns, washing cars without a shut-off nozzle, hosing down sidewalks, or watering lawns within 48 hours of a storm" (James, 2022). They mention that these regulations apply statewide, and there are no exceptions. A home water stewardship guide can give step-by-step instructions on conserving water during droughts. For example, opt for short showers instead of baths and turn on the water only to get wet; when at the laundromat, operate clothes washers only when fully loaded or adjust the water level for the load size; use the automatic dishwasher; choose native plants that use less water and an efficient irrigation system for the lawn. Use a commercial car wash that recycles water. Water-saving techniques may be a possible solution to droughts because saving water from our rivers, bays, and reservoirs help us keep our environment healthy.

In conclusion, we now know that drought affects the environment in many ways. As a high school student, I can produce an action plan for my community. My plan is (a) involve the community representative; (b) learn how droughts have affected us in the past and how they will likely affect us in the future; (c) establish a system to monitor and communicate about drought conditions in the community; (d) prepare and document a set of actions to take before and in response to drought; (e) educate the public; and finally, (f) the most important action is to research other workable solutions against drought and reduce water use. "We are the first generation to feel the sting of climate change, and we are the last generation that can do something about it," (Inslee, 2019). Drought is one of the most damaging natural disasters we have ever faced, and by producing innovative solutions, we can reduce the effects of droughts and implement measures to prevent them.

Bibliography

Balsom, Paul. "Do Water Restrictions Effectively Mitigate Droughts?" *High Tide*, 23 Sept. 2021, https://htt.io/do-water-restrictions-effectively-mitigate-droughts/.

"Interview with Kristen Guirguis."

James, Ian. "California Adopts Drought Rules Outlawing Water Wasting, with Fines of up to $500." *Los Angeles Times*, Los Angeles Times, 4 Jan. 2022, https://www.latimes.com/environment/story/2022-01-04/california-adopts-drought-rules-outlawing-water-wasting.

Notman, Nina. "Atmospheric Water Harvesting." *Chemistry World*, Chemistry World, 7 June 2021, https://www.chemistryworld.com/features/atmospheric-water-harvesting/4011929.article.

"Rainwater Collection." *Centers for Disease Control and Prevention*, Centers for Disease Control and Prevention, 16 Mar. 2021, https://www.cdc.gov/healthywater/drinking/private/rainwater-collection.html.

Robert Sanders, Media relations| August 27, and Robert Sanders. "Water Harvester Makes It Easy to Quench Your Thirst in the Desert." *Berkeley News*, 5 Sept. 2019, https://news.berkeley.edu/2019/08/27/water-harvester-makes-it-easy-to-quench-your-thirst-in-the-desert/.

Vyas, Kashyap. "8 Innovative Drought Solutions That We Can Count 0N." *RIPE*, 16 June 2019, https://ripe.illinois.edu/news/8-innovative-drought-solutions-we-can-count-0n.

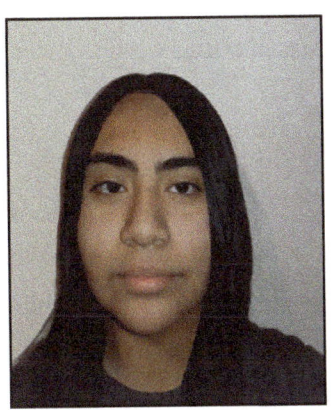

How to Fix Climate Change's Effect in California

Ariana Ortiz

Greenfield High School, Greenfield, CA

Benjamin Franklin once said, "When the well's dry, we know the worth of water." We will feel the consequences if we do not address California's drought. One of the most important climate science concerns in my home state is drought. On the Central Coast, many cities have been experiencing a severe drought that has worsened, affecting the agricultural community. The years 2020, 2021, and the beginning of 2022 have seen severe drought and hot weather. Knowing this can make us aware of how we are causing the water shortage; to stop this, we must focus on saving water. We can certainly make a difference. However, not doing something can lead to larger issues. "California's agriculture has lost over $1 billion (about $3 per person in the US) and more than 14 thousand jobs have been lost due to drought conditions from 2021," according to KVPR, a Valley Public Radio station. California does not only produce enough food to feed the nation but the world. If this drought worsens, it will eventually affect the world. Californians should voluntarily reduce the amount of water usage to improve the drought situation, and more investments should be made for the future.

Better actions must be taken to communicate the seriousness of this issue. California is using more water, not less. A water conservation law in California from 2018 states that a person is only allowed 55 gallons of water per day until 2025 and 50 gallons in 2030; individuals will not be fined for over usage of water. Governor Gavin Newsom has asked urban residents to reduce 15% of water usage compared to 2020. With the goal of 15% in July of 2021, California residents only reduced water usage by 1.8% compared to 2020. Some counties increased water usage. These strategies for solving this problem are not highly effective.

California should invest more in cloud seeding to increase the amount of rainfall. Cloud seeding is one of the most promising ways to fix or reduce the drought caused by climate change. Cloud seeding is done by adding small particles of silver iodide to clouds using aircraft or drones. China has been investing in cloud seeding, bringing more rainfall; it has been remarkably successful. If we were to invest more in cloud seeding, we would see more rain. China's investment in weather modifications has resulted in better air quality. Interesting Engineering mentions, "Since 'artificial rain was the only disruptive event in this period,' it was unlikely that the drop in pollution was due to natural causes." Even if it is not extremely successful, it has shown no negative effects. The government should consider China's success and invest more in cloud seeding. The Hill mentioned, "States like Idaho, Wyoming and Oregon have found success in cloud seeding practices." If these states can successfully use cloud seeding, so can California-America's largest agriculture state. If we put effort into mastering how to produce artificial rain, we can succeed like China. We cannot make it rain all we want; therefore, we must make additional changes to see a difference.

Everyone needs to install water-saving devices in their homes. The less water we use now, the more we will have in the future. Scientific Research, An Academic Publisher, states, "A unit increase in water saving devices will lead to a 0.512 decrease in water consumption level" (Ali et al., 2020). Homeowners should see the benefits of having these water-saving devices; it helps the environment and will save money. How can these installations lower water usage and save money? Energy Gov Au states, "Replacing an old single-flush toilet (12L flush) with a 4-star toilet (3.5L flush) will save 50kL and $148 each year on water bills" (n.d.). Old toilets use more water for one simple flush (12L), while new toilets use less (3.5L). If everyone got a water-saving toilet, we would use about 29% less water when flushing. Also, California should have WaterFix plumbers like in Australia who check homes for water leaks and fittings. Sydney Water states, "minor leaks repaired for free." Saving water is not hard and should be done by everyone, or we will have to face even drier drought conditions.

Israel is known as a leader in water conservation; they recycle a large amount of their water. Recycling water has been proven to reduce pollution; not only that, we can drink it, grow crops, and do so much more. Recycling is a solid plan because we would not constantly be relying on imported water. CalMatters mentioned, "With recycled water, California communities don't have to rely on imported water, which can be cut off during severe droughts or a serious earthquake" (West, 2021). Therefore, Governor Gavin Newsom and the legislature must invest their commitment to recycled water. California only recycles 10% of wastewater; we need to make that 100%. According to Wired, "The goal of Operation NEXT is to upgrade the Hyperion Water Reclamation Plant so that it recycles 100 percent of its wastewater by 2035, producing enough purified water to sustain nearly a million households in LA" (Simon, 2021). Recycling water would not just benefit us but the environment too. Wastewater is a problem in California, polluting lakes, rivers, and the ocean. If our drought conditions were to become more severe, without any investments in recycling water, everyone would feel those consequences because of California's crucial role. CalMatters also mentions, "The current drought is severe, but policymakers and water managers know the situation could be much worse without the previous investments in water recycling." Investments are one of the only ways we can combat severe drought conditions, but they can only work if done in advance.

Drought is not something that is only affecting us because of climate change; every country on Earth is undergoing drought conditions. Climate change is caused by human activity; therefore, we need to fix the consequences of our actions. The government can invest in necessary resources, cloud seeding, and technology to recycle wastewater. However, the government is not the only one that needs to act; we, as California's residents, should voluntarily buy water-saving devices. If the entire world decided to ignore these solutions, the effects would be terrible; we would be desperate for clean water. Being only a high school student, I can make a difference by educating people around me about California's drought conditions because of climate change. Educated people will think twice when wasting water and reduce usage. Reducing water usage is the only way to better the future.

Bibliography

Ali, Shirin. "New Study Says China Controlled Its Weather This Summer." *The Hill*, The Hill, 7 Mar. 2022, https://thehill.com/changing-america/sustainability/climate-change/584502-new-study-says-china-controlled-its-weather/.

Commentary, Guest. "Secure California's Future Water Supply and Invest in Recycled Water." *CalMatters*, 19 July 2021, https://calmatters.org/commentary/2021/07/secure-californias-future-water-supply-and-invest-in-recycled-water/.

Dinneen, James. "Can Cloud Seeding Help Quench the Thirst of the U.S. West?" *Yale E360*, 3 Mar. 2022, https://e360.yale.edu/features/can-cloud-seeding-help-quench-the-thirst-of-the-u.s.-west.

Home, https://www.sydneywater.com.au/.

Journals - Scientific Research Publishing, https://www.scirp.org/journal/index.aspx.

"KVPR - NPR for Central California." *KVPR Valley Public Radio*, https://www.kvpr.org/.

"Main Features - Water Conservation." *Australian Bureau of Statistics*, 3 Oct. 2013, https://www.abs.gov.au/ausstats/abs@.nsf/Lookup/4602.0.55.003main+features5Mar 2013.

Ozdemir, Derya. "China Used Cloud-Seeding to Modify the Weather for Political Celebration." *Interesting Engineering*, Interesting Engineering, 7 Dec. 2021, https://interestingengineering.com/china-used-cloud-seeding-to-modify-the-weather-for-political-celebration.

Simon, Matt. "A Massive Water Recycling Proposal Could Help Ease Drought." *Wired*, Conde Nast, 7 July 2021, https://www.wired.com/story/a-massive-water-recycling-proposal-could-help-ease-drought/.

California Against Drought

America Jasmine Palomarez Ledesma

Greenfield High School, Greenfield, CA

"Can we strike a balance between the comforts of modern life and the health of the planet?" This quote from "The Water Crisis" by National Geographic on YouTube puts into perspective the problem we are facing with our lack of water due to drought.

Drought can happen in hot and dusty areas, but usually, it is described as a lack of rainfall or snow over certain times. With droughts, we risk not having enough water. Droughts have affected areas worldwide, and since the last major drought California experienced (2012-2016), the state has shown an effort to help. Examples include establishing new standards, the Sustainable Groundwater Management Act, and analyzing the drought risk. Today, we will focus on what both individuals and community members can do to make a difference against drought.

Our first strategy is how the government of Australia helps its people through drought season. "Australia passed legislation that allowed the federal government to provide funding to Melbourne for an integrated response to the drought. It also allocated power to a regional water manager to force co-operation between water utilities, city agencies and reservoir managers" from "What Southern Africa Can Learn from Other Countries about Adapting to Drought - World" (Slaughter & Mantel, 2018). We see that governors and policymakers can do a better job of getting involved with their people. Australia helps Melbourne and informs local water managers and water utilities how to help. Australia also motivates people at home to play their part. California's greywater system uses clean water for plant life, which is also a way to help at home. Melbourne has taught us that collaboration and communication are key details for continuing to save water every day.

How we manage our water can affect whether we are wasting it. In "Lessons From Fighting a Drought," N.S Ramnath writes, " . . . coordination among different parties and institutions . . . decide on how and when to use and allocate water . . . technology makes it possible to use water efficiently, to reuse water that would have been wasted, to desalinate brackish water . . . you are conserving water" (2016). We need to start thinking about how we can reduce water usage, and there are many types of technologies that can help, satellites, which help farmers limit water use, and strip drips, which are a service to leaks. To help dissipate fog and weaken storms, cloud-seeding drones and wave-powered desalination systems reclaim water by dissolving any unnecessary salts. Many people can greatly benefit from using this technology, including farmers and community members.

Morocco's Fog Catchers is a method to combat drought. In North Africa, Morocco uses nets that collect fog and transports it in the form of water. As the author claims, water can be reached out of thin air, "Morocco has been utilizing a system of nets that catches fog and converts it to water, in response to climate change and a lack of rainfall in the Southern Moroccan region . . . condensation builds up in the mesh netting and collects in troughs directly under the nets, which is then transported via a piping system" (Chapman, 2019). Working with farmers, we can put different nets around the Central Coast and all-around California. The Fog Collection program at California State University, Monterey Bay (CSUMB) is a nearby project that helps us better understand how this works. On CSUMB's website, the Fernandez Lab analysis says, "The fog collected is a function both of the density of water in the air as well as the wind speed and direction" (2016). Transforming fog into water can be utilized when we are running low; the perfect technology.

Overall, we have seen that drought is complicated but manageable. Improving our preparations for drought season, using technology, and the invention of fog catchers will help. Climate change is a global issue that affects millions of people. As a full-time high school student, I can play my part as an individual and help my area. The biggest way I can help is by reducing my water usage. As we learned, conserving water will leave it for future usage when a drought occurs. Just shutting off the faucet or fixing a leak can save much water, educating my family on the importance of drought can convince them to install low-flow plumbing fixtures, and most importantly, educating myself to become environmentally conscious can be a tool to implement useful habits. What can you do to help your community continue allowing future generations to live with clean water?

Planting trees is another way to help fight against drought. A tree's roots can store water that is soaked up from the ground and send it to the surface, but along with that, planting trees increases biodiversity and absorption of Carbon dioxide from the air, improving the air quality. According to Tamar Shiloh Vidan in an article titled "Planting Forests Can Increase Rainfall and Help Fight Drought in

Europe, Study Shows,""A new study says that converting agricultural land into forests in Europe could boost summer rainfall by an average of 7.6 percent" (2021). Boosting rainfall could be essential to our current drought situation in California. Tree-planting organizations such as A Living Tribute will plant trees on a loved one's behalf. One Tree Planted is also an organization that has planted over 5.4 million trees in California; they provide effortless ways for you to help fight against climate change. Trees are cut down to make paper, build houses, clear space, and for other reasons; consider how easily accessible and beneficial volunteering to plant a tree is and help replenish the Earth.

Action must be taken against California's drought. It is important to take the drought seriously because it can majorly impact people's lives. I can help make a difference by spreading the word about water conservation and encouraging others to make good choices like watering lawns more efficiently and effectively, being cautious of how much water is wasted and contributing to tree planting organizations. California's drought has been going on for years; enough is enough.

Bibliography

Chapman, Wilson. "How Countries Are Confronting Water Shortages." *US News*, 1 July 2019, https://www.usnews.com/news/best-countries/slideshows/countries-considering-different-strategies-to-confront-water-shortages.

Craig, Jeremy. "How to Talk about Climate Change." *Phys.org*, Phys.org, 12 Apr. 2019, https://phys.org/news/2019-04-climate.html.

"Fog Collection Project." *California State University Monterey Bay*, https://csumb.edu/fernandezlab/fog-collection-project/.

Ramnath, N.S. "Lessons from Fighting a Drought." *Mint*, 16 May 2016, https://www.livemint.com/Politics/HrOM4GiBI0vQJpCDfejhkN/Lessons-from-fighting-a-drought.html.

Slaughter, Andrew, and Sukhmani Mantel. "What Southern Africa Can Learn from Other Countries about Adapting to Drought - World." *ReliefWeb*, 4 Feb. 2018, https://reliefweb.int/report/world/what-southern-africa-can-learn-other-countries-about-adapting-drought.

California Drought Concern

Madison Wood

Mira Costa High School, Manhattan Beach, CA

As a young girl goes to fill up her water bottle to cure her parched mouth, she turns the handle of the sink. She continues to turn the handle of her sink, anticipating water shooting out from the spout, but nothing happens. This unfortunate incident could be the fate of many families state-wide if the drought continues to surge throughout California. A drought is a prolonged period of low rainfall, leading to a water shortage. California has been in a severe drought for approximately twenty-two years. Drought is the immediate effect of climate change. Climate change is the long-term increase of weather and temperature patterns. The drought stems from the increasing temperatures, which cause an increase in the rate of evaporation. With less available surface water, vegetation and soil consequently begin to dry out. The California drought is of immense importance for a plethora of reasons.

First, the duration of the drought causes major room for concern. Experts are calling the twenty-two-year drought a *megadrought*. Secondly, droughts cause public health issues. Since there is an immense lack of water, many agricultural farms and businesses utilize whatever water they have access to, which can be a source of contamination of drinking water. Ultimately, the most important part of this awful drought is that officials say we do indeed have enough water if we conserve it correctly. Conservation is only possible with a bit of sacrifice from all citizens, city dwellers to farmers. The drought is so terrifying because to think that all citizens can do something to end the mega-drought but currently are not is extremely detrimental.

The severity of this drought is heightened by climate change as there is a lack of snow. The lack of snow causes a lack of runoff and surface water which is essential for drinking water, livestock, and other essential processes. As a result of climate change, the rainy period has become much more concentrated, which in turn lengthens the dry period of the year. Managing water more effectively involves managing groundwater, reservoirs, and agricultural practices with more efficiency. The water system in California was based on the assumption that around 30% of our water would be naturally preserved as snow in the mountains. As the snow melts due to climate change, it runs into streams and rivers, which lead to reservoirs. However, this design does not cater to current weather patterns as climate change becomes an increasingly large issue.

Agriculture is also a key way to increase California's effective use of water, as agriculture utilizes 40% of California's water. Jay Lund, a Professor of Civil and Environmental Engineering at UC Davis, says that since the water usage of the agricultural industry is so large, it is the agricultural industry that is going to have to cut its usage during a mega-drought to have a real effect. He also says that 90% of a farmer's revenue comes from about half of the land, so farmers could decrease the use of irrigation on the half of the land that does not produce sufficient crops.

Lastly, to ever get out of this drought, California's groundwater conservation needs to be much more efficient. In the past, farmers have used groundwater stored in underground aquifers to irrigate their crops during dry years. This process has caused the sinking of the Central Valley due to overpumping. Scientists, politicians, and community members have been somewhat effective in attempting to salvage this issue. Water managers state-wide have been doing their part to conserve the water that is stored in reservoirs by waiting to let water out of the reservoirs until they see a large storm coming. Politicians have been attempting to rectify this issue by passing laws such as the Sustainable Groundwater Management Act. The Sustainable Groundwater Management Act was passed in 2014 to halt the overpumping of aquifers and to balance out the amount of pumping to equal the amount of recharge into aquifers.

On an individual scale, all citizens are able to help the conservation of water. One way for individuals to conserve water is to apply greywater systems to laundry, showers, or lawns. Lawns are a huge way to conserve water by making lawns drought tolerant. Lawns can become drought tolerant by planting native plant species, which also provide a habitat for native species. In order to solve the problem of the drought, there are long-term solutions that are effective if provided more funding for the issue. One possible solution is large-scale desalination. Desalination of water is the process of removing salt and other minerals from seawater and brackish water for human consumption and use. Desalination is a great solution, but it is extremely expensive and requires a multitude of resources. It would require a large push and a large amount of time to get desalination on a large scale. Also, rainwater harvesting has become increasingly popular. Rainwater harvesting is the collection of runoff from a surface for later use. This saved-up water can then be used in dry conditions. Rainwater harvesting is a very effective way to combat the drought if available to all households.

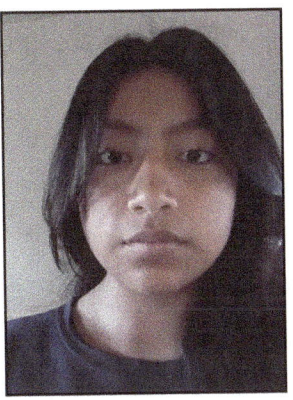

Climate Change

Karina Cruz

Greenfield High School, Greenfield, CA

Elevated temperatures in California cause extreme droughts that cover nearly 87 percent of the state and are a profoundly serious threat to California's agricultural industry, which is responsible for more than half of all domestic fruits and vegetables. Production declines could lead to food shortages and higher prices. As the temperature rises, people start using their air conditioners (AC) more often, which significantly impacts climate change, causing more fossil fuels to be burned, which also causes air pollution. We need to reduce carbon dioxide and use less water, which reduces greenhouse gas emissions.

"When the well is dry, we know the worth of water" Benjamin Franklin. Climate change has greatly impacted California's worsening air quality, affecting people's health. The air has worsened due to hotter and drier conditions, leading to smog, which is composed of air pollutants. People who work outside as the temperature rises increase their risk of injury or death from heatstroke, dehydration, heart attack, and respiratory problems. California experienced a heat wave that led to 140 deaths or more in July 2006. Heat waves could become common by the end of the century.

Reducing fossil fuels would benefit the environment because fossil fuel is one of the reasons why global warming is increasing. As fossil fuels burn, they release copious amounts of carbon dioxide, which traps all the heat in the atmosphere. Fossil fuels create air pollution. Homes use a lot of electricity, and the more we use electricity, the more fossil fuels increase. So, using less electricity would decrease fossil fuels. An article called "Your Guide to Understanding Energy Conservation" by Tara

Energy said, "Energy conservation is about conserving it rather than eliminating it. One of the most effective methods to conserve energy is to turn things down that require a lot of heat or energy" (n.d.). This method shows that people do not have to shut down energy completely but use less. In California, we can conserve energy by shutting down ACs that use freon to cool down the air. "Freon exposure include irritation of the lungs, burns on the esophagus and irritation of the stomach. Necrotic skin lesions or tissue damage may develop when the Freon comes into contact with the body," as stated in the article, "Danger of Breathing Freon by Comfort Solutions HVAC" (2016).

Another way to reduce air pollution is to ride a bike or walk to school instead of riding in a car. Driving a car requires gas and produces emissions, leading to air pollution. So, students walking to school would help the environment. Toyota Boshoku drafted an article called "MONOZUKURI is Overturning the Idea That 'Cars are Bad for the Environment'" which says, "Toxic substances such as carbon dioxide and carbon monoxide are emitted when fuels such as gasoline and diesel oil are burned in automobiles. These substances cause a variety of environmental problems such as air pollution and global warming" (2021). Denmark is a small country that reduced fossil fuels by replacing them with renewable energy. About 50 percent of Denmark's electricity is supplied by wind and solar power. We can reduce fossil fuels by replacing them with renewable energy so that our environment becomes a better place.

In addition, we need to start getting rid of the waste in our atmosphere. Humans create waste, and some waste ends up in landfills. As a result, waste begins to decompose and release harmful gases into the atmosphere, contributing to global warming. Some ways to reduce waste are using reusable water bottles for beverages and reusable plastic bags. A blog by the name "6 Reasons to Blame Plastic Pollution for Climate Change" on the World Bank Blogs website said, "Most people think that when plastic is discarded in recycling bins, it goes away. But there is no "away" – only 9 percent is recycled globally, and the rest is dumped in the natural environment" (Tsydenova & Patil, 2021). This stat means that some of the waste does not disappear like the rest but is dumped in landfills. Recycling prevents emissions of many greenhouse gases and water pollutants and saves energy. An article called "Reduce, Reuse & Recycle" by Boulder County wrote, "Recycling helps to reduce the pollution caused by the extraction and processing of virgin materials. Also, when products are made using recovered rather than virgin materials, less energy is used during manufacturing, and fewer pollutants are emitted" (n.d.).

We also must reduce gas emissions due to the number of people using energy. People use unnecessary energy; to avoid this, we can replace electricity with solar panels. Solar panels use less electricity generated from coal, oil, and gas, which helps protect the environment. The U.S. Energy Information Administration (EIA) article called "Solar Explained" says, "Solar energy technologies and power plants do not produce air pollution or greenhouse gases when operating" (2021). If we start to install solar panels in our homes, it will decrease the chances of gas emissions. Solar panels still work if the sunlight is blocked. If at least 50 percent of our environment installed solar panels, it would decrease the chance of increasing global warming. Replacing electricity with solar panels will help California improve.

Another way to avoid gas emissions is to replace vehicles that use gasoline with electric vehicles. Regular cars produce pollution, which has worsened the environment. An electric car can save an average of 1.5 million grams of carbon compared to regular cars. According to the article called "Benefits of Electric Cars on the Environment" by Électricité de France (EDF), "The major benefit of electric cars is . . . pure electric cars produce no carbon dioxide emissions when driving. This reduces air pollution" (n.d.). Electric cars prevent an increase in air pollution.

Not only can you contribute to climate change outside your home but inside too. Governments worldwide are reducing greenhouse gas emissions by powering economies with renewable energy sources, such as wind or solar farms. Reducing fossil fuels, energy waste, and gas emissions could help our environment decrease global warming. However, if people do not start taking this matter seriously, heat waves will become more frequent, and the environment will not look the same. Therefore, we should take care of this matter right now. As a student from Greenfield Highschool, I want to contribute, so every time I take a shower, I make sure it is not long. I also started taking walks. We need to encourage people and show them what is best for themselves and the environment so that it would become a better place for us and future generations.

Bibliography

Benefits of Electric Cars on the Environment." *EDF*, EDF, https://www.edfenergy.com/for-home/energywise/electric-cars-and-environment.

Comfort Solutions HVAC. "Dangers of Breathing Freon." *Comfort Solutions HVAC*, 5 May 2022, https://www.cshvac.com/dangers-breathing-freon/.

Energy, Tara. "Your Guide to Understanding Energy Conservation." *Tara Energy*, 3 Mar. 2022, https://taraenergy.com/blog/your-guide-to-understanding-energy-conservation/.

"Monozukuri Is Overturning the Idea That 'Cars Are Bad for the Environment.'" *Toyota Boshuku*, Toyota Boshoku Corporation, https://www.toyota-boshoku.com/global/teambreakthrough/technology/001/.

Psydenova, Nina, and Pawan Patil. "6 Reasons to Blame Plastic Pollution for Climate Change." *World Bank Blogs*, 9 Nov. 2021, https://blogs.worldbank.org/endpovertyinsouthasia/6-reasons-blame-plastic-pollution-climate-change.

"Reduce, Reuse & Recycle." *Boulder County*, 21 Apr. 2021, https://www.bouldercounty.org/environment/recycle/reduce-reuse-recycle/.

"U.S. Energy Information Administration - EIA - Independent Statistics and Analysis." *Solar Energy and the Environment - U.S. Energy Information Administration (EIA)*, https://www.eia.gov/energy-explained/solar/solar-energy-and-the-environment.php.

Precipitation Decline - How it is Affecting Our Resources and Environment

Dayra Garcia-Botello

Greenfield High School, Greenfield, CA

California has millions of acres of land, and our home, the Salinas Valley (SV), and the Central Coast (CC) are devoted to planting and farming the food we love. In fact, one nickname for California's SV is *The Salad Bowl of the World* because of its high production of fruits and vegetables; unfortunately, it is affected by droughts and scorching heat. The SV and the CC have always been affected, but only now are we seeing its effects. Dry and empty fields and water shortages are only part of the damage. Farmers have had to leave their farms unplanted from lack of available water, remove trees and plants, have lost millions of dollars to access water, and lost jobs. Other states are forced to grab their produce someplace else due to the slow-down.

What is making one of the richest agricultural states slow its production? One culprit is the heavy decline in precipitation. Rain and snow have always helped farmers keep their land green, but with the 5-year drought of 2012-2016 and wildfires caused by continuous heat, precipitation has been declining faster. Precipitation decline causes droughts, water shortages, and land damage. However, government officials launched the Sustainable Groundwater Management Act (SGMA) in 2014, which changed water laws and pumping in California to slow water pumping and regulate water sources. SGMA also requires community members to waste less water on daily tasks. Although 2022 is still in a drought, this act made huge improvements towards precipitation. However, more needs to be done, and we can unite to overcome these obstacles.

California is no stranger to droughts, a looming problem for our agriculture industry. Since the severe drought of 2012-2016, problems that have not been noticed before have risen; reduced rainfall and snowpacks, constant wildfires, and blazing heat. We have rivers and aqueducts connecting to the CC, but even then, we cannot depend on that. We must reserve more water and reduce current usage. Numerous people use gallons of water for various daily uses, including sprinklers. Sprinklers are not only used to water lawns but also fields. While sprinklers work effectively to water our crops, it is one of the most water-wasting features in agriculture. Instead, I propose that farmers use drip irrigation instead of wide-ranged sprinklers.

Excessive watering leads to extended groundwater pumping, creating many problems for our water sources and fields. According to an Environmental Research Letter titled, "Water Shortage Risks From Perennial Crop Expansion in California's Central Valley," "25% of agricultural regions face intense competition for water resources . . . several have built extensive water storage and conveyance infrastructure" (Mall & Herman, 2019). With rainfall and snowpacks decreasing in larger numbers each year, farmers have resorted to buying or building infrastructures to extract and bring water safely. "Over reliance on these infrastructures can increase economic risks of water shortage" (Srinivasan, 2012). However, drip irrigation spreads water without waste. Water gets into the crops directly and not just around them like sprinklers.

An article written by the University of Rhode Island explained how drip irrigation works, "drip irrigation is 90% efficient at allowing plants to use the water applied . . . Drip irrigation applies water slowly at the plant root zone where it is needed the most" ("Drip Irrigation," n.d.). This technique can be used in fields and help agriculture.

In the Eastern Himalayas, shifting monsoon season (seasonal reversing wind accompanied with precipitation), lack of winter precipitation, water shortages, and unpredictable rainfall hit the region. With this, fears of floods and bursting glacier lakes grew. However, a project launched by World Wildlife (WWF) helped these communities grow crops better. One solution includes "Establishing schools for farmers where they learn how to adapt to climate change with drought-resistant crops and crop rotation" ("Managing Water Scarcity," n.d.). This school for farmers is an example the CC needs to do to preserve water. We must start adapting seeds to grow with less water. What will become of our water sources if we use them irrationally? We need to start producing more drought-resistant plants.

Another solution to save water during this rainfall reduction is creating seed banks to distribute locally. Seed banks are an important aspect of agriculture to prepare for a disaster or climate change and help farmers preserve seeds to prevent the permanent loss of species or plants. "Seed banking is a first fundamental step in implementing the larger road of Biodiversity Initiative..." (Bay Nature, 2019). Seed banks can even help seeds adapt to growth without dependency on water, creating even better farming conditions. In the United States, there are 20 registered seed banks. Once rainfall starts to increase and fields naturally dampen, this safeguarded vegetation will be able to help farmers. For now, however, we must continue working through this emergency.

The local communities on the CC can help overcome water-shortage obstacles by improving home habits and farming techniques in the fields. The rainfall each year is declining, and not slightly. Since this problem is critical enough to be addressed globally, communities, cities, scientists, and governments have chipped in to prevent more shortages. The SGMA helps residents preserve water. Solutions to boost this act's effect include changing how we water crops, growing drought-resistant crops, and creating seed banks to distribute seeds locally. If this precipitation decline crisis is ignored, rainfall will only decrease. Ignoring can lead to consequences we may never want to deal with, reaching farther than just fields.

As a high school student at Greenfield High, the decrease in yearly precipitation affects me daily. Although I cannot change how farmers water their fields overnight, I can do my part to reduce the water wasted at home. I am committed to changing the way we live now. In my school, we can spread awareness and help people notice how our Earth is becoming withered and dry. We can get involved in our community, educate, and bring awareness of just how costly our water is becoming. If we do not do something now, when will we? Are we just going to stand around and wait for it to affect us personally to get concerned? We must change the way we are and how we act. If and only when we do this can we live in a world where a necessity will not become scarce.

Bibliography

Bay Nature Staff. "Nature News: California Funds Seed Banks - Bay Nature Magazine." *Bay Nature*, 16 Sept. 2019, https://baynature.org/article/california-funds-seed-banking-as-hedge-against-climate-change/.

"Drip Irrigation." *URI HomeASyst*, University of Rhode Island, 5 Aug. 2014, https://web.uri.edu/safewater/protecting-water-quality-at-home/sustainable-landscaping/drip-irrigation/.

Hofste, Rutger Willem, et al. "17 Countries, Home to One-Quarter of the World's Population, Face Extremely High Water Stress." *World Resources Institute*, 6 Aug. 2019, https://www.wri.org/insights/17-countries-home-one-quarter-worlds-population-face-extremely-high-water-stress.

Mall, Natalie, and Jonathon D Herman. "Water Shortage Risks from Perennial Crop Expansion in California's Central Valley." *IOP Science*, Environmental Research Letters, 15 Oct. 2019, https://iopscience.iop.org/article/10.1088/1748-9326/ab4035/pdf.

"Managing Water Scarcity." *WWF*, World Wildlife Fund, https://www.worldwildlife.org/projects/managing-water-scarcity.

Spector, Dina. "America's Salad Bowl Is Turning to Dust: Shocking Pictures Show How Bad It's Getting in California." *Business Insider*, Business Insider, 26 May 2014, https://www.businessinsider.com/california-drought-affect-on-food-supply-2014-5#in-the-face-of-water-shortages-many-farmers-have-switched-to-drip-irrigation-which-uses-less-water-than-traditional-sprinklers-18.

Thelwell, Kim. "Seed Banks' Importance for the World." *The Borgen Project*, 7 May 2019, https://borgenproject.org/seed-banks-importance-for-the-world/.

Goodbye Lakehouse

Curran Hedges

Mira Costa High School, Manhattan Beach, CA

"All the water that will ever be is, right now" (National Geographic, 1993).

In California, droughts have always impacted the way we live in a negative way. Sure, we developed some much-needed skills/laws that are beneficial to the sustainability of the human race, but at what cost? The constant battle between man and nature has been going on ever since we first set foot on this rock. We started off taking as much as every other inhabitant and sharing it among other creatures through different methods like the natural cycles. Now, however, we are taking an unimaginable amount of resources, and it is having an enormous impact on the environment and especially our future here. Even though some big public figures refuse to acknowledge the impending doom that climate change imposes, it is a very real issue, and we need to take action. One of the biggest threats of climate change in California is the constant struggle with our drought. Due to our drought, we constantly have to borrow water from other states and limit our water use in certain areas.

The snowball effect is in constant operation here in California. Due to the supply and demand problem of water, we have much more need for water than we can collect on our own. As California gets hotter, we use more water, which we do not have, and end up in a positive feedback loop. The hotter it is, the more water evaporates, as well as the demand for recreational water. If we all gave up our water parks, jacuzzis, 15-minute showers, water fights, and even washing the car in our driveway every

71

couple of weeks, we might not have this big of an issue. However, we cannot just pull the plug on this because not everyone thinks of the environment first. Scientists say this is "the driest 22-year period in at least 1200 years" (James, 2022). The fact that we have done so much harm to our environment in such a short amount of time should prove to everyone that we need to come up with a solution, and a fast one at that. Besides our recreational use being affected by the change in the amount of water, the two biggest issues are the abundance of wildfires and the future of agriculture in California. "With an increase in summer temperatures, the area burned by wildfires has risen fivefold from 1972 to 2018. Warmer summer temperatures and climate-driven aridity are likely to fuel more wildfires in the future" (Scripps Institution of Oceanography, n.d.). With the number of wildfires in the coming years, we will see a change in our campsites, wildlife, agricultural land, and even wildlife reserves/national parks. Another repercussion of this is the health factor. In recent years, we have experienced worsened air quality which correlates to the rising problems with asthma and other respiratory conditions.

It is no secret that greenhouse gas emissions add to climate change, but how much are we adding to it? Even though we are making efforts to reduce emissions coming from California in the long run, we are still faced with the problem of what is happening right now. "If global greenhouse gas emissions continue at current rates, the state is likely to experience further warming by more than 2 degrees F more by 2040, more than 4 degrees F by 2070, and by more than 6 degrees F by 2100" (Scripps Institution of Oceanography, n.d.). It is not the question of what will happen, as we already know what the likely future is for us. It is the question of whether we will do anything to stop it. Unfortunately, even though we have these facts, people refuse to pay attention to them and act as if nothing is wrong.

The real answer is to work together to establish laws that properly protect the environment and stop the situation from getting worse at this fast rate. "The 2022 Scoping Plan Update will assess progress towards achieving the Senate Bill 32's 2030 target and lay out a path to achieve carbon neutrality no later than 2045" (California Air Resources Board, 2021). CARB will initiate the development of modeled scenarios to illustrate outcomes that lead to carbon neutrality (CARB, 2021). This plan will help determine our world's future, especially in California. With these emissions being reduced, climate change will slow down gradually and will help us here in California with our drought issue. "Apart from modest increases in renewable power, the plan relies on carbon capture technology for the elimination of air pollution and restates many of California's existing commitments to slow global warming" (Washington Post). Even though there are more environmentally friendly ways to produce energy, these are not even close to being efficient enough for us to switch right now. There needs to be a lot more research into the efficiency of green energy sources.

There are things we can do at home right now that are beneficial to the environment. The easiest way to do this is to reduce the amount of power and water you use daily. The most important part is spreading awareness of the problem and convincing people around you that they can help and need to, or their way of living will also be impacted. However, our corporations and government can give incentives to help the environment. Not everyone is motivated by moral judgment and being environmentally supportive. I guarantee that if people were rewarded for buying an electric car or reducing power and water consumption, California and the rest of the world would be in a better position. "The earth, the air, the land and the water are not an inheritance from our forefathers but on loan from our children. So we have to handover to them at least as it was handed over to us" (Mahatma Gandhi).

Works Cited

Scripps Institution of Oceanography, "FAQ: Climate Change in California".

https://scripps.ucsd.edu/research/climate-change-resources/faq-climate-change-california

Los Angeles Times, "California just adopted new, tougher water restrictions: What you need to know".

    https://www.latimes.com/environment/story/2022-05-25/california-just-adopted-new-water-re-strictions-what-you-need-to-know#:~:text=Across%20the%20western%20U.S.%2C%20scientists,being%20intensified%20by%20climate%20change.

California Air Resources Board (CARB), "AB 32 Climate Change Scoping Plan".

    https://ww2.arb.ca.gov/our-work/programs/ab-32-climate-change-scoping-plan

The Washington Post, "Tensions rise as drought worsens and heat surges across California".

    https://www.washingtonpost.com/politics/2022/05/23/tensions-rise-drought-worsens-heat-surges-across-california/

Water Scarcity

Alexa Silverio

Greenfield High School. Greenfield, CA

Is there really a climate problem in California? Yes, there is. A big problem on Central Coast, CA, is water scarcity. Water scarcity is a big issue for the Central Coast because it has damaged the agricultural industry. California handles more than half of the domestic food in the U.S. Not only does California feed its people, and the U.S. but some countries too. Water scarcity has a big negative impact on California's citizens. To help fix the water scarcity, officials have declared water restrictions on southern citizens. These restrictions will help people avoid more intense water shortages and will help crops grow without wasting more water. We need to help reduce water scarcity.

Addressing water pollution is a step forward in reducing water scarcity. Water pollution is the contamination of water by chemicals that makes the water unsafe for drinking, cooking, cleaning, swimming, etc. Pollution includes chemicals, trash, bacteria, and parasites. Reducing water pollution is easy and can be done by anyone. Water Scarcity and Quality (unesco.org) has pointed out, "Water quality degradation presents major challenges in securing enough water of good quality to meet human, environmental, social and economic needs to support sustainable development of countries." We need to lessen the pollution in our water to help support our daily lives without having to worry about scarcity. We can generate funding to clean up waste to rid pollution from water. According to California Coastkeeper Alliance (CCKA) CCKA_Citizen-Suit-Report_Final.pdf (cacoastkeeper.org), "Citizen lawsuits not only stop the flow of pollution to California's coast, bays, and rivers, but can also generate funding to clean up pollution and contamination that has occurred." Lowering water pollution can reduce the drought impact, which can also help crops flourish. If water pollution is not reduced, more crops will die; this can affect citizens' lives. How Does Drought Affect Our Lives? | National Drought Mitigation Center (unl.edu) stated, "If those dominoes were drought impacts, the first domino you knock over might be farmers' corn crops dying. The second domino might be that the farmers would not have money to buy a new tractor from the dealer in town. The dealer would then lose money, which would be the third domino. If enough farmers lose their corn crops, the dealership might not be able to employ as many people or may even have to close—the fourth domino. The dealership closing would cause many more impacts in the community." Water pollution is a major problem that we need to fix.

Desalinating more seawater is another solution to help with water scarcity. California already desalinates seawater with 12 seawater desalination facilities that now produce about 12 million gallons (about 45424920 L) per day. If we desalinate more water, it can reduce the amount of fresh water we consume. According to Why desalination could help solve the looming water crisis (sustainablereview.com), "it can provide a low-risk water supply that is not susceptible to natural disasters, drought, or depletion." Desalinating seawater is much safer and more convenient to help droughts and can help California with water scarcity by getting water from a different source that will not affect us. California, parched West remain divided over seawater desalination (usatoday.com) has said, "Desalination can provide a reliable water supply to help address shortages in the region." Seawater will be a great advantage overall to stop water scarcity. If we desalinate more seawater, drought will be much easier to overcome, water scarcity will be much more avoidable, and we might secure our water.

Recycling water can help reduce water scarcity by a lot. Recycled water is water that has been filtered to remove any solids and other chemicals and disinfected by a water treatment plant. Israel has already made 90% of its wastewater recyclable. According to npr.org, "Recycling sewage water has helped free Israel, a desert country, from depending on rain." Water recycling helps Israel overcome their water scarcity and now has reached water security. Water recycling can help California use water to continue to support its people and agricultural industry. California has been restricting its citizens' water supply because of water scarcity, and by recycling water, this issue can be much more controlled. Less damage will be done to the agricultural industry. There will be more water to support California throughout the fire season. Secure California's future water supply and invest in recycled water - CalMatters supports this by saying, "We can safely use recycled water to drink, irrigate parks, support environmental uses, grow crops, produce energy, and much more. More than just a new source of water, water recycling projects supply a degree of local water independence." Water recycling can and will support California and help the Central Coast prepare for droughts. The Central Coast is no stranger to droughts; we can help prevent them from getting worse through water usage and restrictions.

Acting now can reduce water shortage. Officials have declared water restrictions on 6 million people (about twice the population of Arkansas). The restrictions will start on Jun 1. Desalinating more seawater, harvesting rainwater, and recycling water are all practical solutions to reduce shortage problems. If the problem continues, California's agricultural industry will be affected, which is crucial for the U.S. and the world since California makes up more than half of the U.S. food. As a high school student, what can I do? I can help reduce water pollution by using earth-friendly products that do not have microplastic or harsh chemicals and not throwing trash on the ground. I can start cutting unnecessary water from my daily life. Instead of polluting our water, why not try to help reduce water pollution?

Bibliography

Commentary, Guest. "Secure California's Future Water Supply and Invest in Recycled Water." *CalMatters*, 19 July 2021, https://calmatters.org/commentary/2021/07/secure-californias-future-water-supply-and-invest-in-recycled-water/.

"How Does Drought Affect Our Lives?" *UNL*, https://www.drought.unl.edu/Education/DroughtforKids/DroughtEffects.aspx.

NPR, npr.org.

"A SOLUTION TO CALIFORNIA WATER POLLUTION." *California CostKeeper Alliance*, https://cacoastkeeper.org/wp-content/uploads/2018/08/CCKA_Citizen-Suit-Report_Final.pdf.

Sustainable Review. "Why Desalination Could Help Solve the Looming Water Crisis." *Sustainable Review*, 26 Apr. 2021, https://sustainablereview.com/desalination-looming-water-crisis/.

Vasilogambros, Matt. "California and the American West Is Thirsty. but Is Seawater Desalination 'a Silver Bullet' to Solve the Water Crisis?" *USA Today*, Gannett Satellite Information Network, 23 Feb. 2022, https://www.usatoday.com/story/news/nation/2022/02/22/california-seawater-desalination-western-drought/6896322001/.

"Water Scarcity and Quality." *UNESCO*, 3 Dec. 2019, https://en.unesco.org/themes/water-security/hydrology/water-scarcity-and-quality.

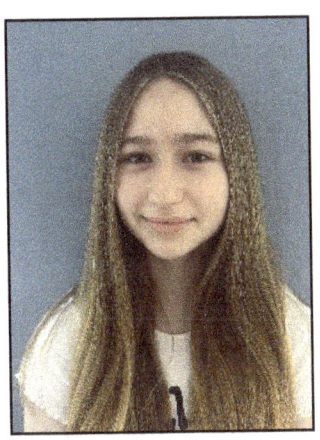

All About California's Drought

Sofia Jordan

Anzar High School, San Benito, CA

California has been in a drought for twenty-two years and counting! A *drought* is defined as a severe lack of precipitation and can be declared after only fifteen days. The US Drought Monitor, or USDM, is a system that is updated every week to show the level of drought in an area. There are five levels of severity, from abnormally dry (D0) to exceptional drought (D4). Did you know that 95.1% of California is in a severe drought (D2), and 59.8% on top of that is in extreme drought (D3) as of mid-May 2022 (USDM, 2022)? Drought is so much more than just not having enough water. Droughts also cause wildfires, increased endangerment to species, economic instability, and even increased human health issues. People should be educated about it, and action should be taken to lessen drought effects.

Severe drought is characterized by a longer fire season with high burn intensity, as well as larger fires. What about extreme droughts? Fire season lasts year-round in extreme drought zones! Even areas that are typically seen as wet areas are at fire risk lately (USDM, 2022). Wildfires can wipe out nature that took decades to develop, place wildlife in harm's way by burning down habitats, and polluting the air. Wildfires also threaten humans and our infrastructure. Humans are affected by air pollution, property and crop loss, and health problems.

Droughts put stress on plants and wildlife, too. A lack of water leads to the soil drying up and animals migrating. Imagine being out on a hot day with nothing to drink, and then imagine how the animals feel living in this environment for weeks on end. Animal populations are decreasing due to the water supply (NCBI, 2012). In correlation, plants do not get adequate water, which ripples through the food web. According to the National Center for Biotechnology Information, "Drought triggered widespread loss of species" (2012). Some animals leave and migrate to areas with more water. The animals that cannot leave oftentimes do not survive because they cannot adapt to the lack of water fast enough. Species that are mostly rural move farther into cities to find food and water. Jason Holley, Supervising Biologist at California's Department of Fish and Wildlife, told Time Magazine, "Those animals that can't adapt aren't going to survive" (Worland, 2015). We are now starting to see the long-term effects of drought on wildlife in California.

Drought affects the economy because it affects agriculture. In 2020, agriculture was responsible for $49.1 billion of California's income (CDFA, 2021). According to the University of California, Merced, "The 2021 drought directly cost the California agriculture sector about $1.1 billion and nearly 8,750 full- and part-time jobs" (Anderson, 2022). In addition, thousands of households in California rely on agriculture as a source of income. Agriculture is more than just a source of food on the dinner table; it is also the root of many families' livelihoods.

Beyond affecting jobs, drought also causes health concerns. Low rainfall means less water flow, which means more stagnant water. Stagnant water, combined with high heat, leads to increased bacteria making water unsafe for humans and wildlife to drink (CDC, 2020). Drought reduces the amount of food you can grow locally. Since water is important for hygiene, drought affects your cleanliness and increases the likelihood of getting some diseases. E. coli and Salmonella are diseases caused by bacteria, which spread easier when water is not available to cleanse food and surfaces properly. Drought, combined with fires, worsens air quality. This increase in smoke, dust and air particles can make respiratory conditions worse and put you at risk for diseases and infections. For example, asthma cases have been increasing yearly, according to the CDC (2011). Overall, drought negatively impacts everyone's health.

Due to all the disastrous impacts of droughts, action is needed. California lawmakers are making changes and suggestions to reduce the drought. Are they doing enough to enforce it? The root of the drought lies in the hands of more than just water usage; it lies in the hands of the way we interact with water as a society. Many people are careless about water use. We need to change the way we water crops. We need to implement drought-tolerant landscapes and green infrastructure. I do not mean green as in color; I mean green as in using nature to help nature. The people of California need to be educated about the far-reaching effects of the drought and how significant they are.

Drought affects everyone and are a problem that everyone needs to help solve. Scientists have come up with multiple potential solutions to reduce the drought. They have developed ways of purifying water, like desalination, and creating smarter ways of managing water, like better irrigation and infrastructure (Cho, 2021). Although solutions exist, they are often less practical and more expensive. Desalination, for example, is a valid solution because it allows vast amounts of salt water to be usable.

However, it is not cost-effective and poses other issues like dealing with the leftover brine. Scientists are busy brainstorming solutions, but society is responsible for implementing them, which has not happened on a wide scale yet. The burden of the drought lies in the hands of businessmen, policymakers, farmers, architects, landscapers, teachers, homeowners, and everyday citizens like you and me.

The drought in California will not end until we make a change. The drought affects wildfires, wildlife, the economy, and our own health! California has been monitoring droughts for 128 years. Surprisingly, 2022 was California's driest year on record! Drought is just one of the many effects of climate change. However, the effects of droughts can be lessened. Everyone can make small changes in their life, and together we can make a difference. How can you make a difference?

Works Cited

Anderson, Lorena, and Josué Medellín. "Last Year's Drought Cost Ag Industry More Than $1 Billion, Thousands of Jobs, New Analysis Shows | Newsroom." *UC Merced News*, 24 February 2022, https://news.ucmerced.edu/news/2022/last-year%E2%80%99s-drought-cost-ag-industry-more-1-billion-thousands-jobs-new-analysis-shows.

"California Agricultural Production Statistics." *California Department of Food and Agriculture*, https://www.cdfa.ca.gov/Statistics/.

"California - Fire Information." *Bureau of Land Management*, https://www.blm.gov/programs/public-safety-and-fire/fire-and-aviation/regional-info/california/fire-restrictions.

"California." *Drought.gov*, https://www.drought.gov/states/california.

Cho, Renee. "A 1,000 Year Drought is Hitting the West. Could Desalination Be a Solution?" *State of the Planet*, 26 August 2021,

https://news.climate.columbia.edu/2021/08/26/a-1000-year-drought-is-hitting-the-west-could-desalination-be-a-solution/.

"Climate change impacts in multispecies systems: drought alters food web size structure in a field experiment." *NCBI*, https://www.ncbi.nlm.nih.gov/pmc/articles/PMC3479754/.

"Health Implications of Drought." *CDC*, https://www.cdc.gov/nceh/drought/implications.htm.

"How Does Drought Affect Our Lives? | National Drought Mitigation Center." *National Drought Mitigation Center*, https://drought.unl.edu/Education/DroughtforKids/DroughtEffects.aspx.

Worland, Justin. "California Drought: Forest and Desert Wildlife Hurt." *Time*, 1 June 2015

https://time.com/3901467/california-drought-wildlife/.

Climate Temperature in California

Noa Benami

Mira Costa High School, Manhattan Beach, CA

Every year, humans play a significant role in making the climate around us warmer over time. As a result, there are many climate science concerns in my home state of California, but one of the most important ones is an increase in global temperatures.

Some people think that a few extra degrees of heat cannot necessarily harm us, but that is not true. At our current rate of increase, there could be long-lasting effects on both the environment and humans. As a form of energy, we burn fossil fuels, which are nonrenewable sources. Nonrenewable means that it cannot be readily replaced at a pace fast enough to keep up with consumption. Therefore, carbon dioxide, methane, and other pollutants are released into the atmosphere. These greenhouse gasses cover the Earth like a blanket, trapping the sun's heat and warming the planet. This result is known as the greenhouse effect, which is the leading cause of rising global temperatures. Average summer temperatures have risen about 3 degrees Fahrenheit since 1896, and detrimental results have been observed by scientists, educators, and the community.

An increase in global climate temperature affects our environment and surroundings in many ways. One example of an impact is a wider area of wildfires and droughts throughout the years. Tree rings show that California is prone to megadroughts that will likely last for decades. With high amounts of heat surrounding us, glaciers melt and cause sea levels to rise, creating flooding and an overflow of water in lakes, rivers, and streams. Coastal communities are often the most impacted by rising sea levels, which includes my community because I live by the beach. Other coastal influences include short heat waves in different spots along the California coast. All the environmental impacts involved with increased global temperatures cause a huge loss in revenue. According to a study, severe storms, floods, droughts, and wildfires caused at least $1 billion in losses in 2015.

An increase in global climate temperature threatens the health of our livelihood and future in many ways. When the Earth's climate becomes warmer, so does the air we breathe. Heat causes water to evaporate faster, creating humidity in hot places. The moisture in the air leads to public health and safety issues involving substances that are dangerous to the human body. Examples of these problems include bad drinking water, hazardous spills, and bad air quality. In addition to impacts from moisture in the air, the recent global temperatures create warmer and wetter waters. As a result, waterborne illnesses and disease-carrying insects can develop. These impacts may not seem significant, but in fact, they have led to more hospitalizations than ever before. Heatstroke, cardiovascular diseases, and kidney diseases are just three of the many health problems that are commonly found.

This concern has been actively discussed between scientists, policymakers, and the general public both in my own community and beyond. With just a quick Google search, there are numerous articles addressing concerns about climate change, how it is caused, and what we can do to help. In addition, many researchers are observing and documenting global temperatures over time. For example, scientists in the Scripps Institute of Oceanography at UC San Diego have been using models to analyze the problem with a series of California climate change assessments. They have even gone to the point of installing 600 cameras to help first responders confirm and monitor wildfires. It is important to continue to spread awareness and display the information so people can make a change.

Making a change is something that comes from us in our own communities. The number one cause of overall climate change is anthropogenic, meaning human influences. If everyone plays their part in reducing their carbon footprint, we can help save our environment before it is too late. So, how can we as individuals play a part in reducing global temperatures? There are many easy and effective ways to save energy and water. First, we can use clean air technology like catalytic converters and smoke scrubbers. Second, we can replace all incandescent lights with LED lights whenever possible. Other steps include switching to renewable resources such as solar and wind power, driving electric cars, installing energy-efficient appliances, and even planting trees.

Now that I have addressed the concerns in my community, I will take action and encourage others to follow and do the same. Even supporting a local business that uses and promotes sustainable climate practices is a great way to begin. I hope others will see the value in reducing increased global temperatures and contribute to making positive changes. It is important to support and care for our local community before it is too late.

Bibliography

"Climate Change in California: Facts, Effects and Solutions." Energy Upgrade California, https://energyupgradeca.org/climate-change

"FAQ: Climate Change in California." Scripps Institution of Oceanography, https://scripps.ucsd.edu/research/climate-change-resources/faq-climate-change-california

Herring, David. "What Can We Do to Slow or Stop Global Warming?" NOAA Climate.gov, https://www.climate.gov/news-features/climate-qa/what-can-we-do-slow-or-stop-global-warming

Melissa Denchak. "Are the Effects of Global Warming Really That Bad?" NRDC, 21 Apr. 2021, https://www.nrdc.org/stories/are-effects-global-warming-really-bad

Photo Credit: Aurora Ellis.

IV.

People * Energy * Persuasion

2022 NATIONAL CLIMATE STUDENT ESSAYIST

Youth Climate Anxiety, Shortsighted Resiliency & Persistent Climate Injustice

Vincent Darius Kreft

The Consequences of a Void in Environmental Policy Leadership
Bloomington High School North, Bloomington, IN

"We can no longer let the people in power decide what hope is. Hope is not passive. Hope is not blah blah blah. Hope is telling the truth. Hope is taking action." – Greta Thunberg

For 26 Conferences of the Parties summits, our world leaders have taken pictures, shaken hands, and made promises, but they have not taken meaningful eco-actions. This lack of climate policy leadership has left the youth of the world anxious. With the lack of investments in climate mitigation, we are now scrambling to enact shortsighted climate resiliency measures that only temporarily protect our communities from climate threats. Unfortunately, these hastily enacted band-aid fixes have often benefited the wealthy while disproportionately placing burdens on our historically marginalized communities. Clearly, our policy leaders need to lead by enacting legislation that corrects both environmental and socio-economic abuse—a carbon tax with earmarked subsidies to our most vulnerable communities could be the answer.

Youth Climate Anxiety

Hickman et al. (2021) conducted a global youth climate anxiety survey with 10,000 respondents across ten countries. The findings were alarming as 84% of global youth (75% of US) were at least moderately worried about climate change. I am a resident of a coal state—according to the US Energy Information Administration, Indiana ranked 3rd in the nation in coal consumption in 2020—so I was curious to replicate the global survey to see if youth in my state are equally anxious. I received 1,349 responses from high school and college students in my hometown of Bloomington, IN, and 80% were at least moderately worried about climate change, five percentage points more than the US.

Diving deeper into the climate anxiety survey results, they show a high correlation between climate distress and perceived inadequacies in government. For example, 63% of US respondents (76% of Indiana) believe that government actions to address climate change are failing young people. Furthermore, 79% of US respondents (80% of Indiana) believe policymakers are not taking climate concerns seriously enough. As highlighted by Greta-inspired climate strikes, there is a growing youth voice that demands more government action (Guterres).

Shortsighted Resiliency and Persistent Climate Injustice

According to *Poverty and Climate Change*, infrastructure improvements can serve as a mechanism to combat climate change (Abeygunawardena et al.). However, poor communities have no relative strength to compete against the lobbying of corporate giants. Those in poverty are lower on the hierarchy of our capitalist society and are constantly bearing disproportionately more environmental burdens. Expanding production facilities, urban sprawl, and our politician's inability to "see beyond the dollar signs" are responsible for our nation's shortsighted inadequacies in mitigating climate change (Rae).

If governments continue to show inaction in passing climate mitigation policies, the disparity between the rich and the poor will grow. Specifically, the rich will utilize their wealth to move to areas of environmental stability (which grow increasingly more expensive and rare), forcing those in poverty to migrate into areas of increasing environmental abuse continuously. Those in poverty will continue to get poorer as the environment gets more threatening, entrenching them in an inhumane quality of life (Comim).

Even Indiana's state capital is guilty of such climate disparity, as rendered in a recent *IndyStar* article. Indianapolis's history of redlining created a cycle of low infrastructural development in communities that were deemed "hazardous." Specifically, poor minority communities were often overlooked for financial support from banks and rich entrepreneurs. As infrastructure became outdated, environmental deterioration soon followed. Although the practice of redlining was banned in 1968, its long-lasting effects are still felt today.

For example, Indianapolis's outdated sewage system cannot handle the modern industrial waste load that now drains into the White River. "One Indiana law not only required industry to locate next to the White River, but it also actually ordered those facilities to dump their waste — harmful chemicals, animal entrails, gasoline — directly into the water" (Bowman, 2021). According to the *IndyStar*, neighborhoods near these industrial overflows have "lower than median incomes, lower property values, and lower life expectancies," (Bowman, 2021). Those in poverty have no options to remove the climate burden placed on them by the companies and government officials that "ultimately passed on those costs" (Parker, 2014). Community-focused government action is needed to stop this cycle of socio-economic injustice (Parker, 2014). If the government utilizes its authority to shift the climate burden off the poor, it would not only rectify climate injustice but would open up more economic opportunities to historically marginalized communities.

Carbon Tax and Climate Justice

Despite the need for government action, policies must not inadvertently increase social abuse through overbearing carbon tax burdens. One risk of a carbon tax is that it exaggerates fuel poverty—specifically, the people already sacrificing significant proportions of their monthly income towards fuel would be hit hardest by a nonflexible carbon tax (Walker). This idea is further explored in a Stanford news article stating that passing a blanket carbon tax is a significant problem due to its perceived unfairness. For example, those with lower income often "drive older, less fuel-efficient cars and frequently commute longer distances to work to access lower-cost housing," which means their livelihood is dependent on using carbon-intensive technology (Parker, 2014). A carbon tax would have to shift the burden of climate change onto the rich rather than continuing to burden these communities already suffering from fuel poverty (Parker, 2014).

According to *Nature Communications*, one way to shift the burden of a carbon tax is to combine it with a subsidy program that targets revenue toward poor communities. Essentially, the rich pay the carbon tax, and then the poor communities receive investments from the money generated by the tax. In other words, economic development and climate justice could be achieved through a carefully constructed carbon tax and targeted reinvestment program (Soergel). Ultimately, poverty and climate change must be conquered simultaneously if our nation wishes to provide an equitable, sustainable, and mentally stable green "American Dream."

Works Cited

Abeygunawardena, Piya; Vyas, Yogesh; Knill, Philipp; Foy, Tim; Harrold, Melissa; Steele, Paul; Tanner, Thomas; Hirsch, Danielle; Oosterman, Maresa; Rooimans, Jaap; Debois, Marc; Lamin, Maria; Liptow, Holger; Mausolf, Elisabeth; Verheyen, Roda; Agrawala, Shardul; Caspary, Georg; Paris, Ramy; Kashyap, Arun; Sharma, Arun; Mathur, Ajay; Sharma, Mahesh; and Sperling, Frank. "Poverty and Climate Change : Reducing the Vulnerability of the Poor through Adaptation." The World Bank. https://documents1.worldbank.org/curated/en/534871468155709473/pdf/521760WP0pover1e-0Box35554B01PUBLIC1.pdf Accessed 10 May 2022.

Bowman, Sarah. "The White River: Boundaries of 'redlining' Maps Still Etched in Indianapolis Neighborhoods." *IndyStar*, Gannett, 2 May 2020,

https://www.indystar.com/in-depth/news/environment/2020/05/02/redlined-indianapolis-areas-still-see-poverty-poor-health/3017810001/

Comim, F. Climate Injustice and Development: A capability perspective. *Development* 51, 344–349 (2008). https://doi.org/10.1057/dev.2008.36

Guterres, A. (2019, March 15). The climate strikers should inspire us all to act at the next UN Summit. https://www.theguardian.com/commentisfree/2019/mar/15/climate-strikers-urgency-un-summit-worldleaders.

Hickman, Caroline; Marks, Elizabeth; Pihkala, Panu; Clayton, Susan; Lewandowski, R. Eric; Mayall, Elouise E.; Wray, Britt; Mellor, Catriona; and van Susteren, Lise. "Climate anxiety in children and young people and their beliefs about government responses to climate change: a global survey," *The Lancet Planetary Health*, Vol. 5: 12, Dec. 2021, pp. e863-e873. https://www.sciencedirect.com/science/article/pii/S2542519621002783

Parker, Clifton B. "Stanford Research Finds Carbon Regulation Burden Heaviest on Poor." Stanford University, 28 Feb. 2014,

http://news.stanford.edu/news/2014/february/kolstad-carbon-tax-022814.html.

Rae, Shonagh. "Editorial: Wealthy Countries Are Responsible for Climate Change, but It's the Poor Who Will Suffer Most." *Los Angeles Times*, 15 Sept. 2019,

https://www.latimes.com/opinion/editorials/la-ed-climate-change-global-warming-part-2-story.html.

Soergel, Bjoern; Kriegler, Elmar; Bondirsky, Benjamin L; Bauer, Nico; Leimbach, Marian; and Popp, Alexander "Combining Ambitious Climate Policies with Efforts to Eradicate Poverty." *Nature Communications*, vol. 12, no. 1, Apr. 2021, p. 2342. www.nature.com, https://doi.org/10.1038/s41467-021-22315-9.

Thunberg, Greta "Blah Blah Blah Speech" #PreCOP26 #Youth4Climate Conference, Milan, Italy (Sept. 2021). https://www.carbonindependent.org/119.html

US Energy Information Administration. *State Profile and Energy Estimates: Indiana* (2021). https://www.eia.gov/state/analysis.php?sid=IN

Walker, Gordon, and Rosie Day. "Fuel Poverty as Injustice: Integrating Distribution, Recognition and Procedure in the Struggle for Affordable Warmth." Energy Policy, vol. 49, Oct. 2012, pp. 69–75. ScienceDirect, https://doi.org/10.1016/j.enpol.2012.01.044.

The Golden State Climate Debate

Carter Dugdale

Mira Costa High School, Manhattan Beach, CA

Over the past century, a sporadic increase in global surface temperatures has threatened many areas worldwide. The South Bay community of Manhattan Beach, California, is no exception, with the sea levels expected to rise over six feet in the next 100 years. This rising sea level extremely concerns the California legislature because there will be the erosion of rocks that houses stand upon, causing them to collapse, along with the intrusion of saltwater into the groundwater. This area is only a tiny portion of all of California, which, as a state, has seen an increase of 3 degrees Fahrenheit in the past century. This increase has caused many problems, such as droughts, heat waves, and health concerns for people susceptible to high temperatures. California's dependence on fossil fuels is the main contributing factor to the increased temperatures. To stop the rise in the future, the Californian government needs to pass climate change laws that will reduce greenhouse emissions. The only ones who can really stop climate change are the people. If everyone did small things such as install solar panels or use reusable energy, greenhouse gases in the air would significantly decrease.

California uses around 10.5 tons of greenhouse gas emissions per person every year. This number has been declining since 2016 with the passage of "SB 32," which strengthened the climate change laws that were already in place. The high emissions of greenhouse gases and other air pollutants in California should be a primary concern for many people because of the effects that come from increased surface temperatures such as increased wildfires, a lack of water, health problems for susceptible people, less snow on the mountains during the winter, and rising sea levels. As the temperature increases, so do the number of fires that take place in the arid areas of California. While climate change does not directly cause the fires, the climate that is created helps to fuel the fires and keep them burning. According to an article by the USDA on the "August Complex" in 2020, the "fires would go on to burn more than 1,032,600 acres, becoming the largest fire in California history" (2021). Without the help of climate change, the fires would have never spread that much. As temperatures increase, the length of periods without rainfall increases as well.

As stated by Laura Anaya-Morga in a Los Angeles Times article, "11.87 inches of rain and snow fell in California in the 2021 water year," which is "half of what experts deem average during a water year in California." This lack of water not only makes the climate drier but also affects agricultural practices and the amount of water used in the farming industry. If there is no rain naturally watering crops, farmers use more water to grow their plants, which leads to an even heavier drought. Another problem with the increasing heat is that it poses serious health problems for people like the elderly and children, who are more vulnerable to respiratory and nervous system problems.

An article that was collectively written by Anna M. Phillips, Tony Barboza, Ruben Vives, and Sean Greene in the Los Angeles Times about the severity of hot temperatures states, "Between 2010 and 2019, the hottest decade on record, California's official data from death certificates attributed 599 deaths . . . but a Times analysis found the true toll is probably six times higher." This quote means that climate change and increased surface temperatures are literally killing people. The deaths of thousands should be enough motivation for people to reduce their greenhouse gas emissions.

In recent years, California legislation has taken a big stance on climate change and reducing greenhouse gases that are released in the state. According to an article posted by the Environmental Defense Fund about the stances of the Californian government on climate change, "After the first decade of AB 32 implementation [(which was a passed legislation that set a limit on the amount of greenhouse gases that are produced)], California's economy is growing while carbon pollution is declining" (n.d.). California has been very active in helping reduce greenhouse gasses that are produced.

The California legislature realized the economic repercussions of climate change, such as human health impacts, property impacts from rising sea levels, and reduced agricultural profit from a lesser amount of water, leading them to take action on the matter. The abovementioned article also provides a portfolio of policies the California legislation will implement soon. The portfolio includes giving incentives to companies to reduce their carbon footprint and expanding "its climate policies, the centerpiece of which is a cap-and-trade program that was extended until 2030" (n.d.). In recent years, the Californian government has taken a leadership role in reducing greenhouse gases, and the passed acts have had majorly beneficial environmental impacts.

None of the policies implemented by the Californian government have any meaning if nobody adheres to them. Everybody needs to do their part in decreasing the greenhouse gasses they emit. Transportation is the main cause of pollution in California, so if everyone used public transportation or an environmentally friendly method such as riding a bike, there would be a reduction in the amount of carbon dioxide in the atmosphere. You can even do something as easy as carpooling to work or school. According to an article by Northwestern entitled, "10 Ways to Stop Global Warming," the easiest and most effective method is "moving your thermostat down just 2 degrees in winter and up 2 degrees in summer [which] could save about 2,000 pounds of carbon dioxide a year" (n.d.). When it comes down to it, the problem is really simple; the little things add up. If everyone does their part and chips in just a little, we will one day live in a cleaner and safer environment.

Works Cited

"Climate Change Could Result in a Beachless Manhattan Beach." *Easy Reader News*, 10 Mar. 2020, https://easyreadernews.com/climate-change-could-result-in-a-beachless-manhattan-beach/.

What Climate Change Means for California - Epa.gov. https://www.epa.gov/sites/default/files/2016-09/documents/climate-change-ca.pdf.

"Climate Change in California: Facts, Effects and Solutions." *Energy Upgrade California*, https://energyupgradeca.org/climate-change.

"California Air Resources Board." *California Greenhouse Gas Emission Inventory Program | California Air Resources Board,* https://ww2.arb.ca.gov/our-work/programs/ghg-inventory-program#:~:text=Per%20capita%20GHG%20emissions%20in,average%20for%20per%20capita%20emissions.

"California Leads Fight to Curb Climate Change." *Environmental Defense Fund,* https://www.edf.org/climate/california-leads-fight-curb-climate-change.

"California Records Driest Year in a Century." *Los Angeles Times*, Los Angeles Times, 18 Oct. 2021, https://www.latimes.com/california/story/2021-10-18/california-records-driest-year-in-a-century.

Mendocino National Forest - Home, https://www.fs.usda.gov/detail/mendocino/home/?cid=FSE-PRD860382#:~:text=The%20fires%20within%20the%20August,largest%20fire%20in%20California%20history.

"Heat Waves Are Far Deadlier than We Think. How California Neglects This Climate Threat." *Los Angeles Times*, Los Angeles Times, https://www.latimes.com/projects/california-extreme-heat-deaths-show-climate-change-risks/.

"Facilities Management, Northwestern University." *10 Ways to Stop Global Warming: Facilities - Northwestern University*, https://www.northwestern.edu/fm/fm-staff/10-ways-to-stop-global-warming.html.

Central Coast California's Climate Change

Maybro Villa

Greenfield High School, Greenfield, CA

The Central Coast of California is among the most rapidly warming regions in the state. A big climate change problem in the Central Coast, CA, is temperature. This issue is most important because a warmer temperature changes weather patterns in a way that dry areas become drier and wet areas become wetter. Scientists, policymakers, and community members have not been effective in their efforts to affect climate change. We need to use more renewable energy and get help from the government to spread awareness.

There needs to be less pollution and more solar energy methods. The less air pollution there is, the less heated it gets outside. According to nationalgeographic.com, "renewable energy sources such as solar and wind do not emit carbon dioxide and other greenhouse gases that contribute to global warming." Less pollution will result from having solar power panels around us; they are a good reusable energy source. Solar panels get their energy from the sunlight and are absorbed by the photovoltaic (PV) cells in the panel. "Renewable energy minimizes carbon pollution and has a much lower impact on our environment." The government can help by providing solar panels, which would lower the possibility of higher climate change. According to ecowatch.com, "all Californians are eligible for the federal solar tax credit, and the state offers several incentive programs and solar rebates aimed at further increasing access to reliable, affordable solar panels" (n.d.). The people in California can pay solar tax credits, and the state will give solar equipment to increase the reliability of affordable solar panels. I think this would be a great way to lower climate change; another way to reduce carbon emissions would be wind energy.

Another example of renewable energy would be wind energy, which has the potential to reduce carbon emissions and the impact of climate change. Wind energy is a relatively small amount of the country's electricity generation, but as technology becomes cheaper, it will become more available. This source is very effective because it supplies about a third of the state's electricity. "Solar and wind generation now account for 23 percent of the California Power Mix." More can be done in our state by installing more wind turbines around counties, cities, and various places. The more wind turbines we have around, the more energy we will get out of it; they create about 8 megawatts of energy per day, which is enough to power six houses for a year. Calwea.org said, "Wind energy projects totaling at least 5,787 megawatts (MW) of capacity are operating in California today, providing enough electricity to power about 2.3 million California households" (n.d.). In addition, wind turbines do not release pollution into the air or water, do not require water to cool down, and are easy to use and effective. "Wind turbines may also reduce the amount of electricity generation from fossil fuels, which results in lower total air pollution and carbon dioxide emissions." Wind turbines are great, but a better way to reduce climate change would be having the government help us.

Establishing policies and incentives can help California transition into a low-carbon, clean energy economy. If we manage to create new laws and new jobs about lowering carbon in the air, we can make people aware of what is happening not just here on the Central Coast but worldwide. I recommend this technique because if we have laws that people have to follow, especially ones that will help the environment, then the chances of dramatic climate change will be low. "In accordance with Section 203 of the Energy Policy Act of 2005 (42 U.S.C. § 15852), each fiscal year the federal government must consume at least 7.5% of its total electricity from renewable sources—referred to as the renewable electricity requirement." We can purchase green power for government operations. Help from the government will better the chances of fighting climate change. Purchasing green power can lower the development of landfill gas energy projects. Landfill Methane Outreach Program (LMOP) helps businesses, states, energy providers, and communities protect the environment and build a sustainable future. "[Green Power Partnership] (GPP) provides resources to states on how they can lead by example by purchasing green power for government operations" (2021).

We can spread awareness to make people understand what is happening worldwide. Having federal action about climate change will provide awareness. By raising awareness, people begin to understand the effects of climate change on health. Awareness will help people have behavioral change and social support for actions needed to reduce pollution in the air, thus impacting climate change. "Raising awareness . . . can also help in getting health-care professionals to support strategies for mitigation and adaptation that will both improve health and reduce vulnerability." Having a larger portion of the population participate is the best way to stop climate change.

Using solar power to slow the process of climate change, wind turbines, and encouraging the government to spread awareness are examples of how we can help stop climate change. However, if we do not do anything to stop climate change and just let it go, we might see the temperature increase by the end of the century. I am a high school student, and I am very committed to lowering climate change because it is important to fix. Now, what might you do to help stop/slow the problem of climate change?

Bibliography

"2018 Total System Electric Generation." *California Energy Commission*, California Energy Commission, 24 June 2019, https://www.energy.ca.gov/data-reports/energy-almanac/california-electricity-data/2020-total-system-electric-generation/2018.

"Fast Facts about California Wind Energy." *CalWEA*, California Wind Energy Association, https://www.calwea.org/fast-facts.

"Homepage - U.S. Energy Information Administration (EIA)." *U.S. Energy Information Administration (EIA)*, U.S. Energy Information Administration, https://www.eia.gov/.

Mccarthy, Gina, et al. "Importance of Renewable Energy in the Fight against Climate Change." *WWF*, World Wildlife Fund, 2015, https://www.worldwildlife.org/magazine/issues/summer-2015/articles/importance-of-renewable-energy-in-the-fight-against-climate-change--3.

Neumeister, Karsten. "2022 California Solar Incentives Guide (Rebates, Tax Credits & More)." *EcoWatch*, EcoWatch, 14 July 2022, https://www.ecowatch.com/solar/incentives/ca.

Nunez, Christina. "Renewable Energy, Facts and Information." *National Geographic*, National Geographic, 3 May 2021, https://www.nationalgeographic.com/environment/article/renewable-energy.

"State Renewable Energy Resources ." *EPA*, Environmental Protection Agency, 5 Nov. 2021, https://www.epa.gov/statelocalenergy/state-renewable-energy-resources.

"WHO/Europe | Home." *World Health Organization*, World Health Organization, https://www.who.int/europe/home?v=welcome.

Climate Change

Carlos Martinez

Greenfield High School, Greenfield, CA

Are we going to sit still and watch our planet die? A climate change concern we have in California is droughts. With more frequent droughts, there will be less water for our agricultural industry. NASA satellites provide water availability to the U.S. Drought Monitor data so farmers can prepare for a drought. We need to think of how we use our resources and generate new resources that will not harm our planet.

Managing Our Water Usage

Knowing the amount of water we are using will help us with how much we really need so that we will not overuse our water resources. Saving water for later use through storage is a great way to manage water resources, as when there is a decrease of water in rivers. "On the Price River, [The Nature Conservancy] TNC has negotiated an innovative water-management agreement with a canal company to enhance flows and agriculture" (2020/2022). They came to an agreement for TNC to manage the water and how it will be used, including the amount of water for agriculture. "The infrastructure that moves irrigation water is often old, and inefficient" (2020/2022). We will need new infrastructures to move irrigation water. Managing our water supply will greatly benefit us.

Using Renewable Energy

Renewable energy does not produce carbon dioxide [see Editor's note]. By using renewable energy, fewer unwanted chemicals will be released on the Earth. "That's because renewable energy sources such as solar and wind don't emit carbon dioxide and other greenhouse gases that contribute to global warming" (Nunez, 2022). By using solar, wind, and other renewable energy sources, you will not be contributing to global warming. Using solar panels as a source of energy will replace other energy sources like fossil fuels that are affecting the environment. "However, installing solar energy systems on land with marginal agricultural value or integrating solar energy systems on farms may provide a variety of economic and environmental benefits to farmers" (U.S. Energy Information Administration, 2022).

Solar energy does not only help us with using fewer fossil fuels and gas emissions, but it also provides benefits for our agriculture industry. Solar energy can be applied with a photovoltaic (PV) panel. PV panels minimize human health issues and use energy effectively. "In addition, new materials, designs, and practices can help to reduce PV manufacturing's environmental impact by minimizing waste, energy use, negative effects on human health and pollution" (Office of Energy Efficiency and Renewable Energy, n.d.). Upgrading PV panels will be most helpful for people. Using renewable energy is a major step toward solving the problem of climate change.

Building homes and roads further away from beaches will allow us and sea life in the wet ecosystem to adapt easier. When the sea level rises, the sea life will not have to struggle to adapt to strange structures. "Many central coast ecosystems are trapped between the ocean and coastal infrastructure (such as roads and buildings), making it difficult for them to adapt as sea level rise" (Myers, 2018). Placing our buildings further away from beaches prevents buildings from being destroyed, which will amount to a loss of money. "That means enticing people to build farther away from the coast, creating a buffer zone of parks and natural areas that can serve dual functions as ecosystems and barriers protecting human development from flooding and storm surge[s]" (Myers, 2018). Building further away from the coast will mutually benefit us and the ecosystem. Removing the old roads and buildings will add more space for beaches to move inland. "We need to restore wetlands, which can help buffer storms and allow for them to move inland with the rising tideline" (Myers, 2018). Restoring wetlands will buffer storms and will reduce more damage. Building new structures further away from coastlines will greatly benefit us and prepare us for sea level rise.

Coming up with a new energy source will benefit our only Earth and us. Climate change affects the world in many ways, such as causing an increase in sea level and temperature. These changes are not good for the planet. The government is reducing gas emissions and creating programs that provide information about clean energy; new energy sources that are not harmful to the planet. We must manage our water resources and how much we use for agriculture. Building structures further away from beaches will help sea life and prevent building destruction from sea level rises which will cost us money. What I can do as a high school student is walk to school or ride a bike. I can persuade other people to do the same. I alone cannot do it all, but with many of us, we will be able to slowly but surely make a change.

Bibliography

"Federal Action on Climate." *Center for Climate and Energy Solutions*, Center for Climate and Energy Solutions, 10 Feb. 2021, https://www.c2es.org/content/federal-action-on-climate/.

Nunez, Christina. "Renewable Energy, Facts and Information." *National Geographic*, National Geographic, 3 May 2021, https://www.nationalgeographic.com/environment/article/renewable-energy.

"Q&A With Monique Myers: How Will Climate Change Affect California's Central Coast?" *California Sea Grant*, California Sea Grant, 21 Aug. 2018, https://caseagrant.ucsd.edu/news/qa-with-monique-myers-how-will-climate-change-affect-californias-central-coast.

"Solar Explained - Solar Energy and the Environment." *U.S. Energy Information Administration (EIA)*, U.S. Energy Information Administration, 25 Feb. 2022, https://www.eia.gov/energyexplained/solar/solar-energy-and-the-environment.php.

"Solutions to Address Water Scarcity in the U.S." *The Nature Conservancy*, The Nature Conservancy, 13 Feb. 2020, https://www.nature.org/en-us/what-we-do/our-priorities/provide-food-and-water-sustainably/food-and-water-stories/solutions-address-water-scarcity-us/.

Publicity and Electric Vehicles in Connecticut

Junseong Jo

Canterbury School, New Milford, CT

For all the publicity electric vehicles (EVs) get for reducing air pollution, less than four percent of EVs were sold in the United States in June, far lower than in China and Europe. So if we all collectively agree that EVs are the way of the future, why aren't more Americans buying them?

The state of Connecticut seems eager to bring EVs to its roads. The Connecticut Electric Vehicle Coalition acknowledges that "low-income and minority communities are often among the worst affected by air pollution," echoing a 2019 study that found that Hispanics and Black people are "burdened with about 60 percent more pollutants, on average, than they cause," meaning that they suffer from the pollution they did not create. Connecticut tries to encourage its residents to purchase EVs through rebate systems. Connecticut Hydrogen and Electric Automobile Purchase Rebate program has rebates of $7,500 incentives for Connecticut residents who purchase EVs. According to lower-income rebate recipients, 56% said that "the rebate was extremely important for their ability to buy ... Sixty-two percent said they wouldn't have purchased it without it." With so many great rebates, one would think Connecticut roads would be swarming with EVs.

However, this is not the case; there are only 21,382 EVs on Connecticut roads. Why is this? In Connecticut, if you want to buy an EV like a Tesla, you must go to Massachusetts, New York, or Rhode Island to purchase the vehicle. Travel is necessary because EV manufacturers sell directly to consumers, which Connecticut's dealer franchise law prohibits. So, despite Connecticut's Public Utilities Regulatory Authority (PURA) issuing the Equitable Modern Grid Initiative, "a nine-year program to support the installation of charging stations across the state," a focus on equity and inclusion, and setting up the "goal of putting 125,000 to 150,000 electric vehicles on Connecticut roads," EVs are not common. Although Connecticut's rebate program impressively "issued more than 5,300 rebates totaling about $10 million since 2015," (Prevost, 2019) most of the rebates applied to buyers who were already well off enough to afford a Tesla.

An incentive or rebate program cannot serve its purpose if a resident cannot access the service. Connecticut senators recognize this and are trying to make EVs more accessible to the customers by passing legislation. For example, Connecticut Democratic Senator Will Haskell tried to pass SB 127, which would have allowed EV companies such as Tesla to directly sell their products to their customers without going through franchised dealerships. Haskell emphasized the importance of SB 127, arguing that "not a lot of good a charging station is going to do if we're not able to easily and conveniently buy electric vehicles in Connecticut."

Bills like SB 127 not only reduce the price for customers but also address the issue of discrimination that minority car owners often face. When Deborah Caviness, the founder of the Southern Connecticut Black Chamber of Commerce, visited a car dealership in Fairfield County to purchase an EV, the salesman became "condescending" after she rejected the dealership's offer. Minority car owners are often forced to pay more than white customers when negotiating to finance, and "Connecticut's existing protections for car buyers … do not go far enough in protecting minority buyers from discrimination."

The passage of bills like SB 127 is essential in bringing EVs to Connecticut roads and urban communities that suffer the negative effects of air pollution. It does not make sense to force a car owner to travel to Massachusetts, Rhode Island, or New York to purchase an EV. A petition is a good idea to spread the awareness and truth about the benefits of direct sales of EVs. William Cross stated that "opposing EV freedom hurts the planet and natural competition," and he strongly encouraged residents in CT to sign the petition so that policymakers would not "listen to dealers no matter how much money they throw at them." This petition will allow CT residents to realize the importance of direct sales, and it will eventually increase the accessibility of EVs in CT. As a high school student in CT, I may also try to publish in local newspapers or post on social media to inform residents about this issue and encourage them to vote for this petition since "It's now or never — we need the votes."

Work Cited

Calma, Justine. "The EV Revolution's next Big Roadblock: Access to Chargers." The Verge, The Verge, 20 Dec. 2021, https://www.theverge.com/22846965/ev-charging-stations-bipartisan-infrastructure-law-equity.

Crider, ByJohnna. "Connecticut Dealerships Are Trying to Block the New EV Freedom Bill - Sign Petition for EV Freedom." CleanTechnica, 9 Apr. 2021, https://cleantechnica.com/2021/04/09/connecticut-dealerships-are-trying-to-block-the-new-ev-freedom-bill-sign-petition-for-ev-freedom/.

The Connecticut Electric Vehicle Coalition. "CT EV Coalition Responds to Deep EV Roadmap." EV Club of CT, 12 Nov. 2019, https://evclubct.com/ct-ev-coalition-responds-to-deep-ev-roadmap/.

Moritz, John. "Advocates for Direct-Sales Say Car Dealership Model Hurts Minority Customers." Connecticut Post, Connecticut Post, 20 Apr. 2022, https://www.ctpost.com/news/article/Advocates-for-direct-sales-say-car-dealership-17111716.php.

Penn, Ivan, et al. "Electric Cars for Everyone? Not Unless They Get Cheaper." The New York Times, The New York Times, 9 Aug. 2021, https://www.nytimes.com/2021/08/09/business/energy-environment/biden-electric-cars-cost.html.

Prevost, Lisa, and Energy News Network November 25 Lisa Prevost. "Connecticut Aims to Put Equity Front and Center in Electric Vehicle Plan." Energy News Network, 22 Nov. 2019, https://energynews.us/2019/11/25/connecticut-aims-to-put-equity-front-and-center-in-electric-vehicle-plan/.

Public Utilities Regulatory Authorities. (n.d.). Pura establishes Statewide Electric Vehicle Charging Program. CT.gov. Retrieved May 27, 2022, from https://portal.ct.gov/PURA/Press-Releases/2021/PURA-Establishes-Statewide-Electric-Vehicle-Charging-Program

Schott, Paul. "Push to Allow Direct Electric Car Sales in Connecticut Revs up Again. Will It Pass?" CT Insider, CTInsider, 8 Feb. 2022, https://www.ctinsider.com/business/article/Push-to-allow-electric-car-sales-in-Connecticut-16840791.php.

Photo Credit: ©Joel Sartore/ Photo Ark

V.

Biodiversity * Floods * Wildfires * Social Justice * Policy Impacts

Racing Extinction: Guide & Education. Oceanic Preservation Society. (2022, August 1).
    https://www.opsociety.org/ourwork/ films/racingextinction

2021 NATIONAL CLIMATE STUDENT ESSAYIST

Biodiversity Loss

Patrick Cannon

Philadelphia High School for the Creative and Performing Arts, Philadelphia, PA

"Look closely at nature. Every species is a masterpiece, exquisitely adapted to the particular environment in which it has survived. Who are we to destroy or even diminish biodiversity?" - E.O. Wilson

Climate change is a global issue impacting different forms of life throughout our planet. It can be connected to various issues such as global warming, rising sea levels, social justice issues, and a multitude of more. It has substantial amounts of impact on our planet's functions in several ways. Amongst these issues, biodiversity loss poses a threat to all life, being a huge issue affecting our planet on a wide scale. Unfortunately, looking out for the best interest of our planet has become politically corrupt, leaving the future unknown.

We are facing a huge climate crisis worldwide. Biodiversity is extremely ecologically important for species in an ecosystem. Biodiversity is the different biotic factors, represented by the variety of life in an ecosystem. In ecosystems, everything is interconnected. This is supported by the Diversity-Stability theory. Within an ecosystem, biodiversity helps with speed recovery from natural disasters, climate stability, ecosystem productivity, and more. Studies have shown from the University of Michigan's U-M Sustainability department, demonstrating that biodiversity affects ecosystems as those comparable to global warming or air pollution. This means, the adverse effects of biodiversity loss are something that needs to be addressed adequately, that will effectively influence action. Biodiversity also is beneficial to humans. Biodiverse ecosystems have numerous positive benefits such as providing food, producing oxygen, cleaning water, and controlling disease. Protection of these ecosystems that are enriched with life and provide so much for humans needs to be prioritized.

Pennsylvania is no exception to the tragic loss of biodiversity throughout the world. It is known that more than 25,000 native species throughout the state are dealing with difficulties due to climate change to survive. Some of these vulnerable animals include two-thirds of bird species inhabiting the state. This includes the Pennsylvania state-bird, the ruffed grouse. Environmental issues globally and within the United States are interconnected. Biodiversity loss is commonly due to different aspects of climate change, including deforestation, global warming, and exploitation of natural resources. Each of these is due to human impact. According to Jim Bonner, the executive director of the Audubon Society of Western Pennsylvania, birds prefer cool environments. Unfortunately, due to the greenhouse gas emissions, causing global temperatures to rise higher than ever before, these birds are facing a dangerous future.

What is being done about biodiversity loss in Pennsylvania? There has been a lack of focus on both the effects of climate change and biodiversity loss throughout the state. Policymakers play a huge part in this, as they commonly are deciding to ignore the impacts environmental issues have. The Pennsylvania Department of Conservation & Natural Resources is an effective work focused on conservation loss through Biodiversity Management, their Wildlife Conservation Program, and Pennsylvania Natural Heritage Program. These programs finance research, conduct inventories regarding conservation work, and conduct management activities. While they effectively use adaptation to manage the damage of climate change, little climate mitigation remains to be made globally. Climate mitigation is focusing on limiting climate change and rising greenhouse gas emissions. To effectively stop climate change and prevent biodiversity loss we need substantial change. Laws that regulate different environmental aspects throughout the planet need to be made. This includes laws prohibiting the destruction of biodiverse forests, limiting greenhouse gas emissions, and more. To efficiently combat climate change, I believe we as a nation must turn to renewable energy.

From the endangered polar bears in the Arctic to the corals in the Great Barrier Reef that are becoming deceased, and even to the Pennsylvania state bird facing endangerment, biodiversity loss is a climate emergency. There needs to be a focus on biodiversity as it's noted to be one of the top five drivers of climate change. Without major changes, the future of our planet and its beautiful biodiversity isn't bright. I strongly believe we need a collaboration of both individual efforts and policy changes to effectively address biodiversity loss. As individuals, some things we can do to promote the protection of species are growing a garden, conserving water use, buying produce locally, and protecting endangered species that are local to you. Another way to take action is by reaching out to policymakers.

Voicing your concerns on climate change could influence potential policy creations. But one of the most effective ways for youth to have impact is through activism. Generation Z can be the generation to influence the prospective changes for the environment. The events that will occur if climate action doesn't take place, will largely impact our future. Which means climate action and the protection of biodiversity is insanely important for our generation, and our biggest weapon is getting our voices heard. In conclusion, biodiversity loss is a detrimental occurrence globally, that needs to be addressed on a wide scale throughout the world.

References

Ad, C. (March 17, 2020) Ruffed Grouse Decline Linked to Loss of Young Forest. Retrieved from https://timberdoodle.org/news/ruffed-grouse-decline-linked-loss-young- forest#:~:text=The%20 decline%20is%20growing%20in,mosquito%2Dborne%20West%20Nile %20virus

Ann, A. (May 2, 2012) Ecosystems effects of biodiversity loss could rival impacts of climate change, pollution. Retrieved from: https://news.umich.edu/ecosystem-effects-of-biodiversityloss-could-rival-impacts-of-climate-change-pollution/

Mary, T. (October 10, 2019) Study: Climate change could make dozens of birds extinct in Pennsylvania. Retrieved from: https://triblive.com/news/pennsylvania/study-climate-change-could-make-dozens-of-bird-species-extinct-inpennsylvania/#:~:text=Pennsylvania%27s%20 state%20bird%2C%20the%20ruffed,Audubon%2 0Society%20study%20released%20Thursday.&text=Last%20month%2C%20the%20journal%20 Science,billion%20birds%20since%20the%20 1970s

Pennsylvania Department of Conservation & Natural Resources. (n.d) Retrieved from: https://www.dcnr.pa.gov/Conservation/Biodiversity/Pages/default.aspx

Elsa, E. (2011) Biodiversity and Ecosystem Stability. Retrieved from: https://www.nature.com/scitable/knowledge/library/biodiversity-and-ecosystem-stability17059965/

An Introduction to Pennsylvania Species, Habitat, Ecosystems, and Biodiversity. (n.d) Retrieved from: (http://www.envirothonpa.org/pdfs/PASpecies_Ecosystems&Biodiversity.pdf

Western Pennsylvania Conservancy. (n.d) Retrieved from: https://waterlandlife.org/wildlifepnhp/species-at-risk-in-pennsylvania/bats/

2020 NATIONAL CLIMATE STUDENT ESSAYIST

Storm clouds and smoke plumes: How climate change disproportionately affects marginalized communities

Stella Carman

Pullman High School, Pullman, WA

We punched through waterlogged drywall with our fists and kicked shelves from walls. Stomping through rotten flooring, we turned devastation into a childish game, the silliness a momentary distraction, replaced by heartbreak as we drove, covered in mud and mold, down streets strewn with wedding photographs, baby clothes, and kitchen tables, the remnants of so many lives now reduced to sloppy muck. Just days before, Hurricane Harvey had ripped apart the gulf coast of Texas, the latest in a series of more frequent and powerful storms. For weeks, my father and I volunteered to help gut some of the badly flooded houses.

As the wreckage in the streets turned from soggy belongings to countertops and cabinets, the gated community in the back of my suburb had a line of garbage trucks at the ready, while the poor, mostly black and Latino neighborhoods had debris crowding the streets for months. The city paid no mind to those who could not afford their own waste services; inner city neighborhoods were left abandoned and further marginalized.

In November of that year, my family moved to Pullman, Washington, a small town on the eastern edge of the state. Months earlier, while Houston drowned in Hurricane Harvey, this small, rural expanse of woodlands and wheat fields made national headlines for having the poorest air quality in the U.S., the result of a summer of record-breaking wildfires and suffocating smoke. Even with the stark contrast between states, their demographics, and the nature of these disasters, it was easy to see that the problems had grown from the same catastrophic phenomenon: global climate change.

Like hurricane season on the East Coast, increasingly violent wildfires have become a fact of summer in the American West. According to a recent study, "anthropogenic contributions to climate change are estimated to have led to a doubling of the total area burned by forest fires in the western US between 1984–2015" (Abatzoglou). Current climate models estimate that the Northwest will lose roughly 1.1 million acres per year to wildfires in the next twenty years (ecology.wa.gov). The economic consequences of such destruction are often the focus of attention when we talk about the problem of climate change, while the impact to human health is largely overlooked. Just like the marginalized communities in Houston during Hurricane Harvey, Washington's indigenous tribes are among those whose health is at highest risk.

Although air pollution levels are decreasing in urban areas across the country, levels of ambient particulate matter, the component of air pollution known to damage the lungs, are rising in the Northwest. This elevation has been directly attributed to the increasing frequency of wildfires. Evidence consistently demonstrates that exacerbations of asthma and chronic obstructive lung disease are associated with wildfire outbreaks. A 2018 study estimates that "the health costs of wildfire smoke exposure range from $11–20 billion/year in the continental US" (Reid).

Historically, public health crises have disproportionately impacted vulnerable populations who often have both limited resources and limited access to health care. Among the most vulnerable in the United States are Native Americans. On average, Native American life expectancy is 5.5 years shorter than the general population. In Washington state, home to 29 federally recognized Native American tribes, the disparity is even more pronounced: the average life expectancy of a Native American male is 69.6, almost ten years below the national average (Dankovchik).

These alarming statistics stem, in part, from the high rates of poverty on reservations, but they are exacerbated by lack of access to health-related education and basic health care. In 2017, the federal budget for Indian Health Services (IHS) was $1,297 per person. In comparison, $6,973 were allotted per inmate in the federal prison system (Whitney). Congress consistently approves lower budgets for IHS, forcing healthcare administrators to cut staff and limit available services. This also means that funding for more or newer facilities is completely inaccessible. So, although respiratory illnesses caused or exacerbated by forest fire are not more common in Native Americans than the general population, the gap in healthcare funding, facilities, and access to care make their occurrence more dire. Once again, those with fewest resources bear the greatest share of the burden.

Although state-wide efforts have been made to address climate change, they have been met with resistance. In 2016 and again in 2018, Washington voted down a proposed carbon tax. While political progress at the state level has been slow, Native American tribal leaders seem to have a greater sense of urgency. In 2019 over 250 people from 41 tribes attended a climate summit in Spokane (Waltower). In October 2020, tribal leaders will reconvene to discuss and develop a policy platform to be used in moving climate action forward in the U.S. and around the world.

As we look to Native leaders and government officials for guidance, we must not make the mistake of believing the problem will be solved by politics alone. We must work to create a widespread culture of environmentally conscientious behavior, one that has the power to influence public policy, corporate practices, and popular attitude. I firmly believe that action must be taken, so in the fall of 2019, I created an environmental club at Pullman High School, not with the hope of creating a tight circle of like-minded students, but as a way to bridge the ideological gaps that stand between us. Only by understanding how deeply the health of our planet connects us, despite state lines and tribal boundaries, can we successfully address the challenges that lie ahead.

Works Cited

Abatzoglou, John T, and A Park Williams. "Impact of anthropogenic climate change on wildfire across western US forests." *Proceedings of the National Academy of Sciences of the United States of America* vol. 113,42 (2016): 11770-11775. doi:10.1073/pnas.1607171113

Climate Change Increases the Risk of Wildfires. Washington State Department of Ecology. https://ecology.wa.gov/Air-Climate/Climate-change/Climate-change-the-environment/Wildfire-risks. Accessed May 25, 2020

Dankovchik, Jenine et al. "Disparities in life expectancy of pacific northwest American Indians and Alaska natives: analysis of linkage-corrected life tables." *Public health reports (Washington, D.C.: 1974)* vol. 130,1 (2015): 71-80. doi:10.1177/003335491513000109

Reid, Colleen E.; Maestas, Melissa May. Wildfire smoke exposure under climate change: impact on respiratory health of affected communities. Current Opinion in Pulmonary Medicine vol. 25,2 (2019): 179-187.

Waltower, Shayna. "40 Native American Tribes Attend Spokane Climate Change Summit" *KREM*. July 30, 2019.

Whitney, Eric. "Native Americans Feel Invisible in U.S. Healthcare System" *NPR*. December 12, 2017.

2019 NATIONAL CLIMATE STUDENT ESSAYIST

Climate Change: A Global Dilemma Impacting West Virginia

Jasmine DeMaria

Brooke High School, Brooke County, WV

Climate change is a global dilemma that is having adverse effects on many aspects of nature. Extreme temperatures and changing precipitation patterns attributed to climate change are two major issues plaguing West Virginia, as well as many other places in the world. However, public figures, including our own president, are sweeping this under the rug in order to keep the public's favor.

Between 1906 and 2016, West Virginia's maximum temperature decreased by 5.3 percent and the minimum temperature increased by 7.7 percent. This change in temperature interrupts the ecosystem's balance and routines. Since fish are cold-blooded, they have a difficult time adjusting to warmer or cooler waters than they are used to. Warmer water also causes reduced dissolved oxygen levels which could be harmful to aquatic organisms. For example, the optimum dissolved oxygen level for West Virginia's state fish, the Brook Trout, is greater than 3 ml/g. The heat is also impacting the migration of several species of animals. Many animals rely on cues such as temperature to know when it is time to migrate. However, climate change is altering these cues and in turn removing animals' natural cues as to when to migrate. Some classes of animals that are dependent on water during their breeding season are also at risk. Jefferson Salamanders, for example, migrate to breeding pools that form after heavy rains. However, these pools may evaporate in the higher heats, leaving the salamander's eggs unprotected from the sun and predators.

Precipitation increased by 2.2 percent between 1918 and 2018. This increased amount of precipitation causes more runoff than can be handled. This causes floods that have many negative impacts on the environment and humanity. These floods destroy buildings, cause mudslides, and unbalance the ecosystem. On the June 24, 2016, West Virginia suffered one of the deadliest floods in the state's history. This flood was brought on by substantial amounts of rain. In White Sulphur Springs, for example, rain totals reached 8.29 inches. In the flooding 23 people died, more than 4,000 buildings were destroyed or damaged, and $53 million in damages was done to bridges and highways. These floods not only displaced humans but animals as well. Deer, birds, fish, snakes, and ground-dwelling creatures were forced out of their habitats due to the dangers and lack of food caused by the rising waters. Many ground-dwelling animals die as their burrows fill with water. Even some fish are transported to unfamiliar waters or stranded on land when floodwaters subside.

Coal is West Virginia's number one export. The burning of coal releases significant amounts of carbon dioxide. In 2016, coal-fired power plants created 24% of the energy-related CO_2 emissions. In 2015, the Obama administration created the Clean Power Plan in an attempt to reduce CO2 emissions from power plants. However, under the Trump administration, the Environmental Protection Agency has created a new plan to replace the Clean Power Plan. It is not for the better, as this new plan aims to only reduce carbon dioxide emissions by 0.7 and 1.5 percent by 2030. While every effort is important, the Clean Power Plan was projected to reduce CO2 emissions by approximately 19 percent in the same time frame. Trump stated that "We've ended the war on beautiful, clean coal and we're putting our coal miners back to work." Despite this claim, the number of people employed by mine operators and contractors is at an all-time low, approximately a thousand fewer than those under the Obama administration. The solar industry alone employs over double that of the coal industry. Why would the president support an industry that has so many negative impacts on the environment? The answer lies in those who funded Trump's campaign. The coal industry as a whole is known to fund the Republican party. Robert Murray, founder of the Murray Energy Corporation and one of Trump's major donors, gave Trump a list of policies that are beneficial to the coal companies. President Trump has followed through with several of these policy changes.

In order to combat climate change, laws have to be put in place to reduce emissions from all sources. Rather than being stopped immediately, coal companies need to be phased out and replaced with clean energy sources. Other programs should be put in place to counteract the damage already done by climate change.

To conclude, climate change has negative impacts on many aspects of life that are not being dealt with properly. If those in power do not take proper actions with this issue, the results could be catastrophic for all inhabitants of this planet.

References

Displaced Wildlife. (n.d.). Retrieved from https://flood.unl.edu/wildlife

Eilperin, J. (2018, August 18). Trump coal plant plan could release hundreds of millions of tons of CO2 into air. Retrieved from https://www.telegram.com/news/20180818/trump-coal-plant-plan-could-release-hundreds-of-millions-of-tons-of-co2-into-air

Extreme Temperature Change: Water Temperatures. (n.d.). Retrieved from http://climatechange.lta.org/climate-impacts/water-temperatures/

Foreign Trade Div. (2019, February 28). State Exports from West Virginia. Retrieved from https://www.census.gov/foreign-trade/statistics/state/data/wv.html

Hewett, F. (2018, August 22). Coal Mining Is a Dying Industry. So Why Does it Play an Out-sized Role in Our Energy Policy? Retrieved from https://www.wbur.org/cognoscenti/2018/08/22/trump-epa-coal-pollution-fred-hewett

How Coal Works. (n.d.). Retrieved from https://www.ucsusa.org/clean-energy/all-about-coal/how-coal-works#bf-toc-5

Jarvis, J. (2018, June 13). Scientists debate human involvement in climate change during panel. Retrieved from https://www.wvnews.com/news/wvnews/scientists-debate-human-involvement-in-climate-change-during-panel/article_5edde257-c0ad-5125-a00a-3b5875ea6a3f.html

Moore, T. T. (2011, May 20). Climate Change and Animal Migration. Retrieved from https://www.lclark.edu/live/files/8522-412moor

Oram, B. (n.d.). Dissolved Oxygen in Water. Retrieved from https://www.water-research.net/index.php/dissovled-oxygen-in-water

Patterson, B. (n.d.). Unpacking The Ways Climate Change is Affecting West Virginia. Retrieved from https://www.wvpublic.org/post/unpacking-ways-climate-change-affecting-west-virginia#stream/0

Patterson, B. (2019, February 4). Coal Comeback? Coal At New Low After Two Years Under Trump. Retrieved from https://www.wvpublic.org/post/coal-comeback-coal-new-low-after-two-years-under-trump#stream/0

Pauley, T. K. (n.d.). Salamanders of West Virginia. Retrieved from http://www.wvdnr.gov/publications/PDFFiles/salamanderbrochure.pdf

Raleigh, R. F. 1982. Habitat suitability index models: Brook trout. U.S. Dept. Int., Fish Wildl. https://www.nwrc.usgs.gov/wdb/pub/hsi/hsi-024.pdf

Share Flood of 2016. (n.d.). Retrieved from https://www.wvencyclopedia.org/articles/2443

The Causes of Climate Change. (2019, May 13). Retrieved from https://climate.nasa.gov/causes/

The Clean Power Plan. (n.d.). Retrieved from https://www.ucsusa.org/our-work/global-warming/reduce-emissions/what-is-the-clean-power-plan

Thompson, A. (2016, June 28). Fires Flare Up Out West While Rains Flood East. Retrieved from http://wxshift.com/news/fires-flare-up-out-west-while-rains-flood-east

What Climate Change Means for West Virginia. (2016, August). Retrieved from https://19january2017snapshot.epa.gov/sites/production/files/2016-09/documents/climate-change-wv.pdf

Acknowledgements

Special thanks to Charles Campuzano, Pacific Region Climate Outreach Coordinator volunteer who is a Manager Systems Engineer for L3Harris in Anaheim, CA. We are truly thankful also for Jennifer Schilz, Books Motivate Volunteer Coordinator in Lamont, CO, who is currently completing a M.S. Masters of Sustainable Development at Regis University, and holds a M.S. Database Technology and Software Engineering degree from Regis University. Your coordination and teamwork helped ensure that students via their teachers, guidance counselors, principals and parents throughout the United States received student registration information for the *2022 National Climate Essay Student Competition*.

American educators of excellence continue to amaze and inspire not only their students, but also those of us who understand that their lessons on life and subject matter content are priceless gifts to be cherished for all time. Books Motivate Foundation salutes your heroic efforts in gaining the respect and trust of your students who understood the importance of submitting climate essays worthy of publication.

Books Motivate Foundation Student Advisory Board member Darren Wang from Hampton High School, Allison Park, PA contributed to the adjudication by the Books Motivate Executive Board of the final cover design used for the *Climate Connection: American Student Voices* book in addition to many volunteer environmental blog articles. To our distinguished panel of judges throughout the U.S. who made the difficult decisions on selections among so many deserving student climate essay entries, thank you. Ternerame Gary, owner of TGary Proofing Services, LLC. provided final copy-editing services. To the exceptional volunteers, sponsors, donor friends and community members throughout our climate education journey, you have made a tremendous difference in enriching our daily lives and future. You deserve and win our applause in working to allow the authors of *Climate Connection: American Student Voices* to be heard.

2022 Executive Board
Books Motivate Foundation